Pickering Emulsion and Derived Materials

Special Issue Editors

To Ngai

Syuji Fujii

Guest Editors
To Ngai
The Chinese University of Hong Kong
China

Syuji Fujii
Osaka Institute of Technology
Japan

Editorial Office
MDPI AG
St. Alban-Anlage 66
Basel, Switzerland

This edition is a reprint of the Special Issue published online in the open access journal *Materials* (ISSN 1996-1944) in 2016 (available at: http://www.mdpi.com/journal/materials/special_issues/pickering_emulsion).

For citation purposes, cite each article independently as indicated on the article page online and as indicated below:

Author 1; Author 2; Author 3 etc. Article title. *Journal Name*. **Year**. Article number/page range.

ISBN 978-3-03842-352-2 (Pbk)
ISBN 978-3-03842-353-9 (PDF)

Table of Contents

About the Guest Editors

To Ngai received his B.S. in chemistry with first class honours at the Chinese University of Hong Kong (CUHK) in 1999. In 2003, he obtained his Ph.D. in chemistry in the same university under the supervision of Professor Chi Wu. He moved to BASF (Ludwigshafen, Germany) in 2003, as a postdoctoral fellow for two years in the Polymer Physics Division under the supervision of Dr. Helmut Auweter and Dr. Sven-Holger Behrens. In July 2005, he moved to Professor Timothy P Lodge's group as a postdoctoral fellow in the University of Minnesota. He joined the Department of Chemistry of the CUHK first as a research assistant professor from 2006–2007, and then was appointed as an assistant professor in January 2008. He was promoted early to tenured associate professor in January 2012. His research interests are in various areas of surface, colloid science and soft materials.

Syuji Fujii received his B.S. degree in 1998 and M.S. degree in 2000 from Kobe University (Japan), where he also received his Ph.D. degree in polymer chemistry under the supervision of Professor Masayoshi Okubo in 2003. His postdoctoral studies were carried out at the University of Sussex (UK) from 2003–2004 and at the University of Sheffield (UK) from 2004–2006 with Professor Steven P. Armes. He joined the Osaka Institute of Technology as a lecturer in 2006 and was then promoted to an associate professor in 2013. His major research interests focus on synthetic polymer chemistry, design and characterization of polymer-based particles, and soft dispersed systems stabilized with the particles (emulsion, foam, liquid marble and dry liquid).

Preface to "Pickering Emulsion and Derived Materials"

Particle-stabilized emulsions, today often referred to as Pickering/Ramsden emulsions, are vital in many fields, including personal care products, foods, pharmaceuticals, and oil recovery. The exploitation of these Pickering emulsions for the manufacture of new functional materials has also recently become the subject of intense investigation. While much progress has been made over the past decade, Pickering emulsion still remains a rich topic since many aspects of their behavior have yet to be investigated.

The present "Pickering Emulsion and Derived Materials" Special Issue aims to bring together research and review papers pertaining to the recent developments in the design, fabrication, and application of Pickering emulsions. The content covers the basic principles of colloidal particles confined at liquid/liquid interface, interfacial assembly and emulsion stabilization, as well as using Pickering emulsion as a template to fabricate functional materials for different applications.

We believe this Special Issue will serve as a platform for researchers to share their exciting works and, in the long run, this will attract further attention and interest in the innovative development of this important and promising area.

We would like to gratefully acknowledge our colleagues and friends who have contributed with passion and expertise to this book. In addition, our thanks go to the editorial team for their assistance in preparing this Special Issue.

To Ngai and Syuji Fujii
Guest Editors

materials

MDPI

Review

Controlling Pickering Emulsion Destabilisation: A Route to Fabricating New Materials by Phase Inversion

Catherine P. Whitby [1,*] and Erica J. Wanless [2]

1   Institute of Fundamental Sciences, University of Massey, Palmerston North 4410, New Zealand
2   Priority Research Centre for Advanced Particle Processing and Transport, University of Newcastle, Callaghan, NSW 2308, Australia; Erica.Wanless@newcastle.edu.au
*   Correspondence: c.p.whitby@massey.ac.nz; Tel.: +64-6-951-9007

Academic Editor: Syuji Fujii
Received: 29 June 2016; Accepted: 22 July 2016; Published: 27 July 2016

Abstract: The aim of this paper is to review the key findings about how particle-stabilised (or Pickering) emulsions respond to stress and break down. Over the last ten years, new insights have been gained into how particles attached to droplet (and bubble) surfaces alter the destabilisation mechanisms in emulsions. The conditions under which chemical demulsifiers displace, or detach, particles from the interface were established. Mass transfer between drops and the continuous phase was shown to disrupt the layers of particles attached to drop surfaces. The criteria for causing coalescence by applying physical stress (shear or compression) to Pickering emulsions were characterised. These findings are being used to design the structures of materials formed by breaking Pickering emulsions.

Keywords: Pickering emulsion; particle-stabilised emulsion; destabilisation

1. Introduction

Controlling emulsion stability during their storage and use is a major challenge [1–4]. Emulsions are used in cosmetic products, detergents and foods, as well as for liquid extractions and oil recovery [1–3]. Products like moisturizer creams take advantage of how emulsions yield and flow, their texture and visual appearance. These properties depend on the volume fraction and size of the droplets in the emulsion. They change over time due to Ostwald ripening, flocculation and coalescence of the drops.

Making an emulsion that will not age irreversibly while it is being handled requires addition of stabilising components. They can be surfactant molecules, polymers, proteins, or particles. The topic of this review is emulsion stability in the presence of particles. Ramsden [5] (and later Pickering [6]) first described the presence of a membrane of solid particles (proteins or other precipitated colloids) enhancing the lifetime of oil droplets and air bubbles in water. Recent progress has improved our understanding of the mechanisms by which solid particles slow Ostwald ripening and coalescence [4,7,8].

The remarkable stability of Pickering emulsions is a problem for applications that require controlled destabilisation of emulsions. Particle-stabilised emulsions that form during the extraction of bitumen from oil sands, for example, are difficult to break. They reduce the volume of oil recovered and generate waste (the unwanted emulsion) [9–11]. Pickering emulsion formation during biphasic reactions catalysed by nanoparticles increases the reaction yield, but reduces its efficiency [12,13]. They hinder separation of the products from the reaction mixture and recycling of the catalyst [14,15]. Particle separations using biphasic extractions are also hindered by particles becoming trapped at the liquid interface [16,17].

Emulsions and foams are destabilised to make coatings and adhesives by evaporating the volatile components to leave a film of active ingredients on a solid surface. Using Pickering emulsions as precursors for assembling films of particles on surfaces, for example, relies on the particle-coated drops or bubbles coalescing with a flat oil-water or air-water interface [18,19]. Although Pickering emulsions are templates for assembling particles into porous solids [20–22], the particle networks tend to collapse during drying [23,24]. The voids in the solids formed typically lack the desired polyhedral geometry.

The focus of this review is on developments in our understanding of how particle-stabilised emulsions break down. Denkov et al. [25] first proposed that Pickering emulsions destabilise if there are defects, like fractures or vacancies, in the particle layer coating the drops. The defects cause the thin films separating drops to rupture and the drops coalesce. One approach to controlling Pickering emulsion stability is to synthesise particles that respond to an external stimulus by detaching from the drop surface. The particles are used to form emulsions that can be destabilised on demand. This approach was recently reviewed comprehensively [26]. Here we focus on the structural changes that occur in Pickering emulsions as they age. We describe how emulsions are broken by particle detachment, mass transfer and drop coalescence. Then we discuss the materials being fabricated by harnessing destabilisation processes in Pickering emulsions.

2. Detaching Particles from Fluid Interfaces

Particles (of radius r_p) assemble at oil-water interfaces (of interfacial tension, γ_{ow}) by becoming partially immersed in both liquids and forming a three phase oil-water-particle contact angle, θ_{ow}. Attached particles reduce the total interfacial area between the oil and water. This alters the free energy of the system by changing the balance of surface energies. For spherical particles with $\theta_{ow} < 90°$, the free energy of attaching a particle to a drop is denoted $\Delta_a G$, and is given by [27,28]:

$$\Delta_a G = -\pi \gamma_{ow} r_p^2 \left(1 - cos\theta_{ow}\right)^2 \tag{1}$$

The interfacial energy trapping a particle at an oil-water interface calculated using Eqn 1 is significant for particles on the colloidal length scale (up to $10^4 \ kT$ for r_p ~10 nm). The lack of thermodynamic stability in Pickering emulsions arises because bare oil-water interface remains between the particles attached to the drop surfaces. The positive free energy of forming this interface always outweighs the negative contribution from particle attachment. Although attached particles are not at equilibrium, particles with radii larger than several nanometres attach to oil-water interfaces effectively irreversibly, giving emulsions kinetic stability. In this section we discuss destabilising emulsions by altering the particle wettability or by competitive displacement with surfactant molecules and by modifying the particle flocculation (Figure 1).

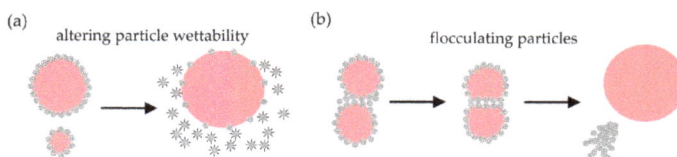

Figure 1. Pickering emulsions are destabilised by detaching particles from the emulsion drops by (**a**) using surfactants that adsorb to the particle surfaces and modify their wettability so that they favour being fully wetted by the continuous phase; or (**b**) enhancing particle flocculation sufficiently to favour adhesion between drops and rupture of the interfacial particle layers which produces flocs of particles and uncoated drops of the dispersed phase.

2.1. Altering Particle Wettability

Attached particles can be displaced by using surfactants that adsorb to the particle surfaces and modify their wettability in situ (Figure 1a) [29–31]. Alargova et al. [29] caused particles to detach from foams stabilised by hydrophobic polymer microrods by gently adding a few drops of a concentrated anionic surfactant (sodium dodecyl sulfate, SDS) solution. They argued that adsorption of SDS onto the particle surfaces made them hydrophilic and caused them to detach from the air–water interface. As the foam collapsed, the particles drained from the foam into the lower aqueous phase. Subramanian et al. [31] observed that isolated bubbles (~100 μm in diameter) stabilised by micrometre-sized latex particles became unstable to disproportionation and particle detachment after exposure to SDS or Triton X-100 (non-ionic surfactant). They also argued that at high surfactant concentrations, the surfactant adsorbed onto the particle surfaces and made them hydrophilic.

How the wettability of micrometre-sized polystyrene latex particles at a planar decane-water interface is altered by adding SDS to the water was investigated by Reynaert et al. [30]. They suggested that altering θ_{ow}, and hence the extent to which the particles are immersed in each liquid, will affect the interaction forces between particles at the interface. Their examination of the particle arrangement in the interfacial layer revealed that the latex particles assembled into network structures that were looser (more open) in the presence of SDS [30]. They showed that the contact angle of a drop of water on a polystyrene film immersed in decane increased as the concentration of SDS increased in the water and argued that this was due to adsorption of SDS [30]. Moreover they argued that adsorption of SDS onto latex particles attached to a decane-water interface must increase the oil wettability of the particles and hence weaken the lateral capillary interaction forces between the particles.

2.2. Competitive Displacement of Particles

Another strategy for displacing particles from drop surfaces is to add surfactant which competes for the oil-water interface [32–34]. Vella et al. [35] observed that adding surfactant could disrupt particle layers attached to planar water surfaces. They examined densely packed monolayers of polymeric particles (r_p = 50 μm) on the surfaces of water-glycerol solutions. They used a needle to inject a drop of non-ionic surfactant (polyoxyethylene sorbitan monoleate) into the layer. Providing the particles were not jammed together, Vella et al. [35] observed a crack form, where the needle touched the particles, and propagate through the monolayer. They argued that localised reduction of the surface tension caused tensile stress in the particle layer, forcing the particles to rearrange [35]. As the crack propagated, the particles consolidated and exposed the liquid surface. Once the particles jammed, they trapped the crack in its final shape for several hours.

Vashisth et al. [34] showed that mixing dodecane-in-water emulsions stabilised by silanised fumed silica nanoparticles (r_p ~10 nm) with solutions of anionic surfactant (SDS) causes the nanoparticles to be displaced from the drop surfaces. Rather than adsorbing onto the particle surfaces (which are already coated with hydrocarbons), the surfactant adsorbs competitively at the oil-water interface. Two minutes of mechanical mixing was required to completely displace the particles from the drop surfaces after adding surfactant at concentrations above the critical micelle concentration [34]. Examination of drop surfaces in emulsions mixed with lower surfactant concentrations revealed that the drops were coated with a mixed layer of nanoparticles and surfactant (Figure 2) [34]. They speculated that particle displacement occurs by a mechanism similar to the displacement of proteins [36] by surfactants. Like particles, proteins stabilise interfaces by forming an immobile, viscoelastic film. Adding surfactant to protein-stabilised emulsions and stirring can induce displacement of proteins from the drop surfaces. This was linked to the reduction of the interfacial tension caused by surfactant adsorption [37].

Katepalli et al. [33] argued that surfactant addition will cause particle displacement from a drop (of radius, r_d) if a surfactant-stabilised drop (of the same radius) has a lower free energy than the particle-stabilised drop. Thus adding surfactant causes particle displacement due to the emulsion system seeking a lower energy state. Katepalli et al. [33] found that for the surfactant-stabilised emulsion to be more stable, the following inequality must be satisfied:

$$\frac{\gamma_{ow} - \gamma_{os}}{\gamma_{ow}} > f \left[\frac{1 - cos\theta_{ow}}{sin\theta_{ow}} \right]^2 \qquad (2)$$

where γ_{os} is the interfacial tension of a surfactant-stabilised drop and f is the fraction of the interfacial area occupied by particles. For particles of intermediate (or neutral wettability ($\theta_{ow} = 90°$), surfactant addition can cause particle displacement if the fractional change in the oil-water interfacial tension with surfactant adsorption is greater than the fraction of the drop surfaces coated with particles. Katepalli et al. [33] examined the response of octane-in-water emulsions stabilised by carbon black nanoparticles to exposure to solutions of surfactant at their critical micelle concentrations. They found that adding Triton X-100, which reduces the interfacial tension by 94% from 51 to 3 mN·m^{-1}, was sufficient to displace carbon black particles from the octane drops [33]. In contrast, adding sodium octyl sulfate only reduced the interfacial tension by 67%, which was not sufficient to cause particle displacement.

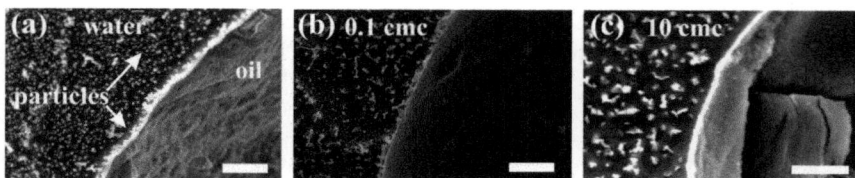

Figure 2. Electron microscopy images of the oil-water interface in a Pickering emulsion mixed with surfactant at concentrations of (**a**) 0 M; (**b**) 10% of the critical micelle concentration; and (**c**) ten times the critical micelle concentration [34]. The layer of densely packed particles at the drop surfaces is disrupted by surfactant adsorption at concentrations below the critical micelle concentration. The particles are completely displaced from the interface at surfactant concentrations above the critical micelle concentration. The scale bars correspond to 2 μm in the left and middle images and 1 μm in image on the right. Adapted from [34] with permission from Elsevier.

2.3. Flocculating Drops and Particles

Particle-stabilised emulsions are also sensitive to flocculation of the particles. Briggs [38] and Lucassen Reynders and van den Tempel [39] first reported that weakly flocculated particles are most efficient at (kinetically) stabilising emulsions. Briggs [38] argued that this was due to strong flocculation producing particle aggregates which are too large to assemble into layers at the surfaces of drops.

Horozov and Binks [40] showed that there is the potential for particle-coated drops to flocculate by particle bridging, where particles attach simultaneously to two drop surfaces. They proposed that bridging occurs where there are strong repulsive interactions between the particles and they form dilute monolayers at the oil-water interface [40]. Bridging occurs when particles on opposing interfaces interlock as the interfaces come together [41,42]. French et al. [43] demonstrated that for particle bridging to occur, there must be insufficient particles present to stabilise the interfacial area in the emulsion and the particles must be preferentially wet by the continuous phase.

Binks and Lumsdon [44] found that drops flocculate in emulsions formed under conditions corresponding to the onset of particle flocculation. Horozov et al. [45] proposed that the drops form three dimensional networks with the particles at the onset of particle flocculation. Evidence of network formation in Pickering foams was found by Chuanuwatanakul et al. [46]. They observed that foams stabilised by mixtures of nanoparticles and surfactants had a granular morphology at surfactant concentrations sufficient to cause strong flocculation of the particles [46]. Subsequently Whitby et al. [47] showed that the energy of adhesion between the particle layers coating the drops increases as the extent of particle flocculation increases. Coalescence is favoured under conditions of strong particle flocculation, where the adhesive energy between the particles is comparable to the energy required to detach the particles from the drops (Figure 1b) [47].

3. Transferring Mass between the Liquid Phases

Pickering emulsions can destabilise by the transfer of mass between drops of different sizes, or between drops and the continuous phase (Figure 3). The former process is known as Ostwald ripening and causes a fraction of the drops to increase in size. In the latter process the average drop volume is reduced (or shrunk) by causing the liquid in the drops to dissolve or evaporate.

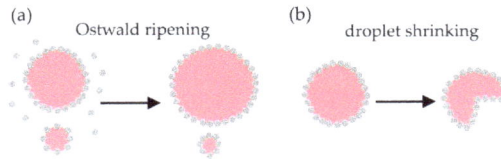

Figure 3. Pickering emulsions destabilise by mass transport due to (**a**) Ostwald ripening with small drops shrinking and larger drops swelling as molecules transfer from the small to the larger drops; and (**b**) all drops shrinking due to the liquid phases in the emulsion becoming partially miscible. The layer of particles coating shrinking drops buckle when the particles become jammed.

3.1. Ostwald Ripening

One of the mechanisms by which drops coarsen in oil-in-water Pickering emulsions (and air-in-water foams) is Ostwald ripening. It is controlled by the molecular solubility of the oil (or air) in the aqueous phase. Molecules in small drops transfer to larger drops due to the difference in Laplace pressure across the fluid interfaces of differently sized drops (Figure 3a). Particles attached to the interfaces in emulsions and foams can arrest Ostwald ripening. Ashby and Binks [48] found that Ostwald ripening in toluene-in-water emulsions formed in the presence of laponite nanoparticles is initially rapid and then ceases at long times. They proposed that particles in the continuous phase may attach to the freshly created surfaces of the growing drops, as illustrated in Figure 3a. Since the clay particles attached to the drops are not easily displaced, they compress into an insoluble barrier around the shrinking drops and eventually halt ripening [48]. Cates [49] argued that Pickering emulsions resist Ostwald ripening if the particles are jammed around the surfaces of shrinking drops, because they cannot follow the drop surface inwards. Instead the interface develops in such a way as to have zero mean curvature. The Laplace pressure thus drops to zero and Ostwald ripening is halted.

That bubbles coated with micrometre-sized latex particles deform over time into polyhedral shapes, with flattened faces and rounded edges and corners was shown by Abkarian et al. [50]. They used simulations to show that these bubble shapes are stable to disproportionation, since this is a minimum energy configuration and the Laplace pressure across the flattened fluid interface is negligible [50]. Meinders and van Vliet [51] used numerical simulations to show that Ostwald ripening in a Pickering emulsion is arrested if particles attached to the drops cause their surfaces to resist compression, and the surface elastic compression modulus (E) is at least twice the surface tension. Later Cervantes Martinez et al. [52] showed experimentally that the condition for stability to Ostwald ripening in a particle-stabilised foam is that $E > \gamma_{aw}/2$.

Attached particles can fail to arrest Ostwald ripening. Ettelaie and Murray [53,54] argued that the rate of bubble dissolution in foams can be faster than the rate at which particles are transported to bubbles and attach to their surfaces. They calculated that the bubble size distribution broadens with time in the case where the particle concentration is higher than that required to stabilise the total air-water interface, since it is governed by the time taken for particle attachment [53,54]. For cases where the particle concentration is not sufficient to stabilise the air-water interface, the bubble size distribution narrows, as the final interfacial area is determined by the number of particles available [53].

The rate of Ostwald ripening in emulsions is slower than the rate of particle attachment to the drop surfaces. In the case where there are insufficient particles to stabilise the total oil-water interface in emulsions, Avendano Juarez and Whitby [55] showed experimentally that destabilisation initially occurs by a combination of droplet flocculation and ripening. Close contact between the flocculated drops enhances oil transfer from smaller drops to larger ones [55]. Large drops swell over time until the density of attached particles is insufficient to protect the drops against coalescence [55].

When two different o/w emulsions containing mutually miscible oils are mixed, mass transfer between the droplets can produce a single population of drops containing a mixture of the oils. This process is called compositional ripening. It is related to Ostwald ripening, however the chemical potential difference due to the concentration differences normally outweighs that due to Laplace pressure differences, and mass transfer is dominated by compositional ripening. Binks et al. [56] showed that compositional ripening in mixtures of Pickering emulsions triggers droplet coalescence, unlike in surfactant-stabilised emulsions where the drops swell, but do not coalesce. They found that adding excess particles suppressed the swelling-triggered coalescence as the particles attach to and stabilise the fresh oil-water interface being created [56]. If coalescence was not suppressed, the merging drops tended to become trapped in non-spherical shapes (this is known as arrested coalescence behavior and is discussed later).

3.2. Shrinking Drops

Where the liquid phases used to form an emulsion become partially miscible, the emulsion can destabilise by droplet shrinking (Figure 3b). Many pairs of immiscible liquids, for example, begin to mix when heated or cooled. Clegg et al. [57] used confocal fluorescence microscopy to visualise drop shrinking in particle-stabilised oil-in-alcohol emulsions as they were slowly warmed up to the temperature where the alcohol and oil formed a single liquid phase (the upper critical solution temperature). They observed that the particle-laden drop surfaces buckled and cracked, and argued that the cracks allow the liquid inside the drops to leave and mix with the external liquid [57].

The shrinking of macroscopic, pendant drops of water in silicone oil that were coated with hydrophobic silica crytals (r_p ~6.5 μm, θ_{ow} = 125°) was visualised by Asekomhe et al. [58]. As water was sucked out of the drops, they lost their spherical shape and buckled [58]. Datta et al. [59] visualised the changes in shape of particle-coated drops in emulsions where the internal phase was slightly soluble in the external phase. They found that an increasing proportion of the drops buckle as the drop volume was systematically reduced. Larger drops buckled more easily than smaller drops [59]. The shrunken drops resembled buckled structures formed by solid shells under compressive stress. These observations supported their hypothesis that densely-packed layers of colloidal particles at fluid interfaces act collectively like solid layers [59]. By measuring the pressure in particle-coated droplets as they were deflated, Xu et al. [60] demonstrated that there is a transition from fluid-like to solid-like behaviour in the particle shell as it is compressed, as shown in Figure 4.

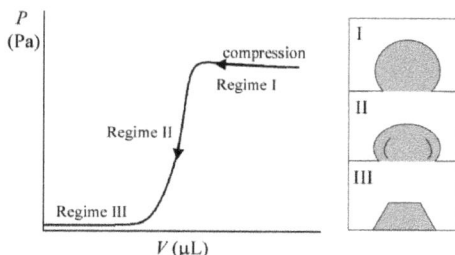

Figure 4. Behaviour of a drop coated with particles as the drop volume is reduced [60]. The pressure inside the drop increases slightly for only small reductions in the drop volume (Regime I). The interface remains fluid-like and the drop profile shrinks isotropically. In Regime II, the particles pack closely together and jam. The pressure inside the drop falls to zero and the drop takes on the shape of a wrinkled hemisphere. At even larger reductions in drop volume (Regime III), the pressure is insensitive to the compression. The particle layer coating the drop buckles and flattens at the top of the drop.

Aveyard and co-workers [61,62] examined the compression of layers of particles at the planar air-water surface of a Langmuir trough. They demonstrated that when a layer of particles at a planar fluid interface is compressed, it bends and forms an undulating surface with a characteristic wavelength. Wrinkling occurs because the area occupied by the attached particles remains constant although the area of the trough has decreased. Lateral compression causes the coating to expand (wrinkle) in the perpendicular directions. Following the general theory for wrinkling of elastic sheets, Vella et al. [63] showed that the periodicity (λ) of the wrinkles in particle coatings can be estimated [63] by:

$$\lambda = \pi \left[\frac{4}{3(1 - \varphi)(1 + \nu)} \right]^{1/4} \sqrt{L_c r_p} \tag{3}$$

where φ is the area fraction of particles at the liquid surface, ν is the Poisson ratio of the particle shell and L_c is the capillary ratio of the drop. Whitby et al. [64] found that the wavelength in crumpled particle layers on coalesced drops is consistent with that predicted by Equation (3).

Razavi et al. [65] investigated the mechanisms by which planar liquid surfaces laden with particles collapse as they are compressed. They used a Langmuir trough to study the surface pressure of air-water surfaces coated with close-packed monolayers of silica spheres ($r_p = 500$ nm) modified to different extents by reaction with dichlorodimethylsilane. Relatively hydrophilic particles formed a fluid-like monolayer that experienced an irreversible collapse. Microscopy images of the surface revealed that this was likely due to expulsion of the particles into the water [65]. In contrast, more hydrophobic particles formed a solid-like, cohesive monolayer that exhibited a prominent compressional elasticity through reversible wrinkling and folding [65]. Stress relaxation was arrested by some of the hydrophobic particles ejecting into the aqueous subphase. Razavi et al. [65] argued that particle-laden oil-water surfaces might show different collapse behavior, due to the long-range forces [66] between the remaining particles that are mediated by the oil phase. Garbin et al. [67] visualised the contraction of pendant oil drops coated with gold nanoparticles in water. The nanoparticles detached from the surface as the drop volume decreased and the particles became close-packed. Furthermore, these workers suggested that short-range steric repulsions between the ligand-capped particles played a crucial role in detachment.

4. Coalescing Drops

Coalescence is a process in which two drops merge to form a larger drop. It reduces the total interfacial area in the emulsion. Coalescence occurs even during emulsion formation and must be (temporarily) halted to impart kinetic stability to an emulsion. Figure 5 shows that the strong (irreversible) attachment of particles to interfaces means that coalescence can be limited or arrested in Pickering emulsions, unlike in surfactant-stabilised foams or emulsions.

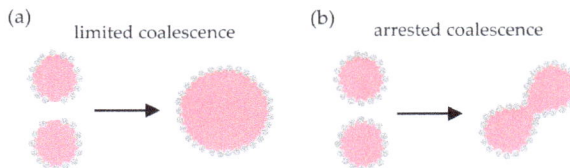

(a) limited coalescence (b) arrested coalescence

Figure 5. Pickering emulsions destabilise by (**a**) limited coalescence until a critical surface coverage of particles is reached if there are not sufficient particles present to fully coat all the drop surfaces that form initially; and by (**b**) arrested coalescence where the combined surface area occupied by particles is higher than the interfacial area that would form by complete coalescence of the drops, so the coalescing drops remain trapped in an intermediate stage of coalescence, unable to relax into a spherical shape.

4.1. Limited Coalescence During Emulsion Formation

Arditty et al. [68] found that if, during emulsion formation, the total number of particles present is not sufficient to fully coat the oil-water interface, the drops will coalesce together until a critical degree of surface coverage by the particles is reached (Figure 5a). By following the evolution of the drop size distribution in emulsions of millimetre-sized drops with a digital camera, they showed that the transient and final drop size distributions are relatively narrow. For a given degree of coverage of the drop surfaces by particles, τ, there is a linear relationship [68] between the average drop radius at any time ($r_d(t)$) and the mass of particles (m_p) which is given by:

$$\frac{1}{r_d(t)} = \frac{s_f m_p}{\tau 3 V_d} \tag{4}$$

where s_f is the droplet surface area covered per unit mass of particles and V_d is the volume of the dispersed phase. This is a generalisation of the relation originally proposed by Wiley [69] to account for observations that the coalescence rate in emulsions containing finely divided solids decreases to zero as the drops approach a limiting size and a relatively uniform size distribution. Fritgers et al. [70] used numerical simulations of the final states formed by a mixture of immiscible fluids and particles to confirm the dependence between the particle concentration and the average drop radius in Pickering emulsions.

Daware and Basavaraj [71] showed that the limited coalescence model can be used to predict the drop size in emulsions formed by using micrometre-sized silica rods to arrest the temperature-induced phase separation of critical mixtures of 2,6 lutidine and water. The model has also been applied to w/o emulsions stabilised by asphaltenes. Pauchard et al. [72] observed that water drops stabilised by asphaltenes will grow until a critical mass coverage of the drop surfaces is reached and argued that this is reminiscent of limited coalescence in Pickering emulsions. They proposed that asphaltenes behave like nanoparticles and jam together into a dense, glassy monolayer at the drop surfaces [72,73].

The limited coalescence model assumes that all the particles in an emulsion are equally effective at stabilising drops. This may not be the case, however, for mixtures of different particles. In the case of oppositely charged particles, it is necessary for the particles to heteroaggregate into networks or clusters to stabilise emulsions. Whitby et al. [74] studied emulsions formed in the presence of mixtures of oppositely charged titania and silica nanoparticles. The titania particles were partially hydrophobic and, on their own, attached strongly to the oil–water interface and stabilised emulsions. The silica particles had hydrophilic surfaces and were poor emulsifiers. Adding silica particles to the titania dispersions enhanced coalescence processes during emulsion formation. This was demonstrated by Whitby et al. [74] using cryogenic scanning electron microscopy to visualise the drop surfaces. They linked the destabilisation to the presence of silica particles in the particle layers at the drop surfaces. Nallamilli et al. [75] modified the limited coalescence model to describe emulsions containing a binary mixture of oppositely charged particles. They successfully predicted the drop size dependence on the number ratio of particles in the mixed system [75].

4.2. Coalescence of Partially-Coated Drops after Emulsion Formation

Emulsions can form with droplets that are only partially covered by particles. Strongly repulsive colloidal particles, which form ordered, dilute planar monolayers at liquid interfaces, can act as effective emulsion stabilisers. The particles bridge the thin films between the drops in close contact. French et al. [43] showed that shearing dilute emulsions of fully-coated oil drops in water could cause particle bridging, by creating more oil-water interfacial area than could be stabilised by the available particles. Similarly, Zhang et al. [76] observed that bridging occurred in emulsions formed by ultrasonication, rather than vortex mixing, due to the larger amount of agitation creating a larger oil-water interfacial area. These emulsions become sensitive to coalescence when the repulsive interactions between the particles are enhanced, or the drops are concentrated together.

Xu et al. [77] found that hexadecane drops formed in aqueous dispersions of polydopamine particles (r_p = 192 nm) were only partially coated by particles at acidic pH. Rather than forming a continuous layer, the polydopamine particles occupied segregated regions on the drop surfaces. They showed that lowering the pH in the emulsions caused the drops to coalesce together and form densely coated drops [77].

Traditional methods of emulsification make it difficult to systematically vary the density of nanoparticles at drop surfaces, as they involve simultaneously forming and fragmenting drops in the presence of particles. For a fixed energy input, the final drop size will vary with the particle concentration, while the particle density at the interface remains constant [68]. Microchannel emulsification offers the advantage of forming monodisperse drops where the size and particle loading of the drops are independently controlled [78]. Drop formation at a flow focusing nozzle occurs so quickly, that the drops must be equilibrated with particles to allow time for attachment to occur. Priest et al. [78] found that dodecane drops formed in microchannels and incompletely covered by silanised fumed silica particles coalesced when they were concentrated together. Similarly, Manga et al. [79] showed that emulsions are unstable to coalescence when drops are formed using rotational membranes without allowing sufficient time for particle attachment to occur.

Fan and Striolo [80] used dissipative particle dynamics simulations to study coalescence of oil-in-water and water-in-oil drops as the density of nanoparticles on their surface was varied. The maximum force and the corresponding drop separation in the force-distance profiles during coalescence were taken as the threshold for coalescence. Their analysis [80] identified the conditions under which coalescence occurred between drops that were partially coated by nanoparticles. These included when the nanoparticles were poorly wetted by the continuous phase, and when the nanoparticles were strongly attracted to each other, or to the approaching drop surface.

4.3. Coalescence Dynamics

Ata [81,82] developed a powerful method for visualising the coalescence dynamics of air bubbles coated with particles. She filmed bubbles at the tips of adjacent capillaries as they were allowed to grow until they came into contact. Pristine (uncoated) bubbles (and drops) coalescing together first oscillate, alternately expanding in the horizontal and vertical directions, until settling into a spherical shape, as shown in Figure 6a [83,84]. Ata [81] found that pristine air bubbles (r_d = 1 mm) oscillated together for more than 40 ms. By analysing films of the projected area of bubbles coated with glass beads (r_p ~33 μm) in the presence of a cationic surfactant (cetyltrimethylammonium bromide), Ata showed that the bubble oscillations were damped within ~30 ms (Figure 6b,c). Ata [82] proposed that the particles reduced the oscillation frequency by forming a semi-rigid shell around the bubble which increased the inertia of the bubble surface. By comparing the coalescence dynamics of bubbles coated with silanised glass beads of different hydrophobicity, Ata and co-workers [85,86] found that attaching particles with higher air-water contact angles made the bubble surfaces more rigid.

Detachment of particles from the bubble surfaces during coalescence of bubbles coated with glass beads (r_p ~33 μm) in the presence of a cationic surfactant was also reported by Ata [82]. This detachment tended to occur at low surfactant concentrations, where the particles were presumably only weakly attached to the surface. Tan et al. [87] investigated particle detachment from bubbles with glass beads that were modified by reactions with silanes or esters to make them hydrophobic. It was predicted that increasing the particle hydrophobicity should reduce particle detachment. The fraction of particles ejected from the bubble surfaces did not vary, however, with the particle hydrophobicity [87]. Tan et al. [87] argued that this implies that particle detachment is dominated by the kinetic energy of the surface oscillations.

Ata et al. [85] observed that coating air bubbles with a close-packed monolayer of latex (r_p = 190 nm) or anatase (r_p = 100 nm) nanoparticles caused damping of the oscillations between coalescing bubbles within ~20 ms. The amplitude of the bubble oscillations was larger than those observed with bubbles coated with glass beads (r_p ~33 μm [81]). This suggested that attached nanoparticles increase the bubble stiffness, but not to the same extent as observed for particles that are tens of micrometres in size [85]. Thompson et al. [88] found that cross-linking nanoparticles at the surfaces of oil drops caused a significant increase in drop resistance to coalescence.

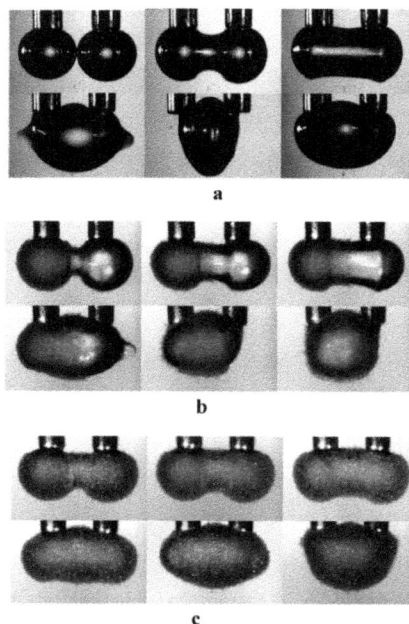

Figure 6. Coalescence behaviour of pairs of bubbles of equal volume generated at the tips of adjacent capillaries (with diameters of 1.07 mm). The time between the photographs is about 0.5 ms. (**a**) Bubbles with pristine surfaces initially oscillate, as the neck that forms between them extends as an expansion wave along the surface of the merging bubble. The merged bubble relaxes into a spherical shape within a few milliseconds; (**b**) In the case where one bubble is coated with silanised glass beads (with a median radius of 33 µm), the oscillations of the merging bubbles drive the beads into the centre of the bubble surface. They form a belt around the bubble surface, leaving each of the ends uncoated. The amplitude of the oscillations gradually decreases until the merged bubble reaches a stable stationary state; (**c**) Two fully coated bubbles coalesce together smoothly. The oscillations of the merging bubble are damped, presumably due to the attached particles making the bubble surface relatively rigid. Reprinted with permission from [81]. Copyright (2008) American Chemical Society.

4.4. Inducing Coalescence by Shear or Compressive Stress

Emulsion stability to coalescence is governed by the thin liquid films of continuous phase locked between touching drops. For a thin film to be in mechanical equilibrium, the repulsive surface forces in the film must balance the external forces pushing the drop surfaces against each other. Rupture occurs once liquid has drained out of the film sufficiently for the surfaces to come close enough for van der Waals attractions to dominate. In the case of Pickering emulsions, the energy required to detach the particles from the drop surfaces is an important contribution to the efficiency with which particles stabilise the drops (see Equation (1)). It will not, however, prevent liquid from draining out of the films. The capillary pressure arising from the deformation of the liquid interface around the attached particles as liquid is squeezed out causes the film to resist thinning and rupture. Thus causing coalescence in emulsions almost always requires deformation of drops by shear or compressive stresses.

Deformation creates extra surface that is not occupied by particles. Rupture can then occur at these "weak spots" in the film.

The response of individual drops to shear or compressive stress reveals the dynamics of drop deformation [89]. Becu and Benyahia [90] studied the deformation and relaxation of individual particle-coated drops under jumps in strain imposed by a counter-rotating shearing device. Retraction of particle-coated drops was slower than the pristine drops, with the drops taking about 20 times longer to return to a spherical shape once the strain was removed [90]. Tan et al. [91] found that individual oil droplets coated with kaolinite particles and compressed by a colloidal particle (that was much larger than the drop) were mechanically robust and recovered their spherical shapes after large deformations. Russell and co-workers [92,93] investigated the effect of covalently cross-linking particles into membranes around drops. They found that cross-linked drops deform irreversibly. While the elasticity of unmodified capsules indicated that the interfacial tension did not change with the applied strain, the elastic response of the cross-linked capsules changed as the strain increased, suggesting the membranes had fractured [92]. Asare-Asher et al. [94] found that water marbles (water drops coated with hydrophobic particles that are only slightly immersed in the water) can withstand deformations of up to 30% and recover their spherical shapes. Higher deformations cause the particle layer to crack [94].

Dilute emulsions tend to yield and flow in response to shear stress. Whitby et al. [95] investigated coalescence in dilute oil-in-water Pickering emulsions (at a drop volume fraction, $\varphi = 0.5$ under shear applied by a rotational rheometer. The bromohexadecane drops ($r_d = 35$ μm) were stabilised by silanised fumed silica particles (r_p ~10 nm) that formed layers a few hundred nanometres thick around the drops. There were excess silica nanoparticles in the water which formed networks that entrapped the drops and enhanced emulsion stability at rest. At dilute φ, coalescence requires drops to collide and remain in contact long enough for the thin film formed between them to drain and rupture. They found that the susceptibility of the drops to orthokinetic coalescence depended on the extent of particle flocculation in the network of particle-coated drops and excess particles [95]. Moreover, they were able to successfully manipulate the particle flocculation and hence the extent of demulsification by varying the salt concentration in the aqueous phase.

Kruglyakov et al. [96] investigated coalescence in dilute Pickering emulsions ($\varphi = 0.5$ as they were compressed in a centrifugal field. The decane-in-water emulsions were stabilised by hydrophilic silica nanoparticles that had been modified by adsorbing cationic surfactant onto their surfaces. They found that the critical capillary pressure required to break the emulsions was significantly lower than the theoretically predicted maximum capillary pressure in a film stabilised by two layers of closely packed spherical particles [96].

Tcholakova et al. [4] compared the critical pressure leading to coalescence measured by Kruglyakov et al. [96] to the critical pressure values calculated for various types of emulsifiers from the available literature data on centrifugation studies of emulsion stability. They scaled the critical pressure by the inverse of the average drop radii in the emulsions. Tcholakova et al. [4] calculated that the scaled value of the critical pressure for surfactant and protein-stabilised emulsions ranges between 0.1 and 0.3 Pa·m. They estimated that the scaled value of the critical pressure measured by Kruglyakov et al. [96] was about an order of magnitude lower (assuming r_d ~20 μm).

Concentrated Pickering emulsions can behave like solids, showing striking rigidity in response to small applied stresses. Arditty et al. [97] found that the elastic storage moduli of Pickering emulsions were significantly higher than those of surfactant-stabilised emulsions. They proposed that strong lateral attractions between attached particles make the drop surfaces extremely rigid. Arditty et al. [97] tested the coalescence stability of the emulsions by centrifuging the emulsions. They found that the magnitude of the compressive stress required to induce coalescence was consistent with that predicted using the elastic coefficient of the drop surfaces. The scaled value of the critical pressure measured by Arditty et al. [97] was about 0.3 Pa·m.

Hermes and Clegg [98] investigated the yielding behaviour of concentrated emulsions of oil drops ($r_d = 7.5$ μm) in water stabilised by silica nanoparticles ($r_p = 330$ nm). Adding high concentrations of salt (44.8 wt. %) caused the drops to flocculate and they saw evidence for this being due to aggregation between particles attached to neighbouring drops. The flocculated emulsion flowed once sufficient strain was applied to break the drop clusters apart. Compressing the emulsion to drop volume fractions, φ ~0.95 increased its elasticity by two orders of magnitude. At these φ, the rate-determining step for coalescence is rupture of the thin films between drops. The drops in the compressed emulsion coalesced instead of flowing when the emulsion yielded. Hermes and Clegg [98] argued that this was due to the particle stabilisation of the interface failing at high strains. The scaled value of the critical pressure was about 0.2 Pa· m.

4.5. Partial and Arrested Coalescence

Partial coalescence of drops can occur in o/w emulsions where the oil drops contain crystals, if a few crystals protrude out from the drops into the continuous water phase. When one drop collides with another, the protruding crystals can pierce the thin water film between the drops and be wetted by the oil in the other drop [99–101]. Walstra and co-workers [99,100] found that if there is sufficient liquid oil in the drops, the oil will flow around the crystal. The crystals within the drops form a solid network [102,103] that hinders the relaxation of the globules into a spherical shape. The partially coalesced droplets form an irregular shape (sometimes called a clump) [99,100]. Thivilliers-Arvis et al. [104] showed that the rate of partial coalescence depends on the size of the crystals and the extent to which they protrude from the drop surfaces.

Partial coalescence can also be observed in o/w emulsions containing solid particles that are completely wetted by the oil. Frostad et al. [105] measured the interaction forces between two solid-in-oil-droplets attached to two capillaries immersed in water. The magnitude of the forces measured between drops as they came into close contact were consistent with those predicted assuming that capillary bridges form between the drops [105]. Pawar et al. [106] observed coalescence between partially crystalline oil droplets, as the elastic modulus of the oil was varied by altering the concentration of fat crystals in the drops. At low particle concentrations, the drops behaved like weak gels with low elastic moduli. During coalescence the interfacial energy dominated the elastic energy and the merged drop relaxed into a spherical shape. The drop elastic modulus increased as the solids fraction increased. Pawar et al. [106] showed that at a critical particle concentration, the particle network elasticity balanced the Laplace driving force and prevented the merging drop from relaxing into a spherical shape.

Studart et al. [107] showed that when monodisperse drops partially coated with particles (formed using microchannel emulsification) are closely packed together they will coalesce into stable, non-spherical structures of two or more drops. They used electron microscopy imaging to show that there were dense layers of particles jammed together over the surface of the merged drops [107]. The attached particle layers on the merging drops make their surfaces sufficiently viscoelastic to increase the characteristic time for shape relaxation and (effectively) arrest coalescence. They argued that coalescence at one site on a drop surface does not lead to particle jamming across the whole oil-water interface. Providing patches of pristine drop surface remained exposed, a drop could undergo arrested coalescence with more than one adjacent droplet.

Pawar et al. [108] used a micromanipulation technique to make in situ observations of coalescence events between pairs of drops, of the same volume, each with a precisely known (fractional) surface coverage by particles (C_1 and C_2). They observed that total coalescence dominates for initial surface coverage values of $C_1 + C_2 < 1.43$ [108]. Pairs of drops, each with a fractional surface coverage of 0.9, were stable to coalescence. Arrested coalescence was favoured at $1.43 < C_1 + C_2 < 1.8$. Under these conditions, the combined surface area occupied by particles was higher than the interfacial area that would form by complete coalescence of the drops (Figure 5b) [108].

Morse et al. [109] visualised arrested coalescence between pairs of oil drops partially coated by polymer nanoparticles, providing cross-linker was present in one of the drops. They found that holding the drops in contact for about 60 s, and then decompressing them, led to arrested coalescence. Morse et al. [109] argued that the interfacial particles in the contact area become cross-linked and that moving the drops apart disrupts the cross-linked particles sufficiently to expose pristine drop surface and promote coalescence between the drops. Deformation of the drop surfaces caused by the separation of the drops may have also contributed [110].

Whitby et al. [111] found evidence of arrested coalescence in bulk Pickering emulsions. They used confocal fluorescence microscopy to visualise droplet packing in o/w emulsions stabilised by silanised silica particles as the emulsions were compressed. At the volume fraction where the emulsions started to break down, the drops increased in size, with some forming arrested shapes, as shown in Figure 7 [111]. Tan et al. [112] showed that arrested coalescence in emulsions can be triggered by Ostwald ripening of nanoparticles coating the drop surfaces. They studied paraffin-in-water emulsions coated with freshly precipitated $Mg(OH)_2$ nanoparticles. Emulsions stored at temperatures higher than 80 °C destabilised [112]. A fraction of the drop population coalesced into non-spherical drop shapes. They argued that large precipitated nanoparticles grow in size at the expense of smaller nanoparticles at the elevated temperatures. This reduces the density of the particle coverage on the drop surfaces and triggers arrested coalescence of the drops.

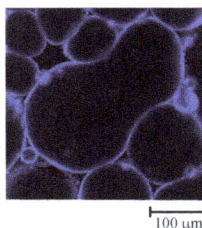

Figure 7. Confocal fluorescence image of the particles shells in an unstable, compressed Pickering emulsion showing a coalesced droplet fused into a doublet shape. Adapted from [111] with permission from the Royal Society of Chemistry.

Whitby and Krebsz [64] investigated the rheology of concentrated oil-in-water emulsions stabilised by silanised fumed silica nanoparticles. These silica particles were shown to aggregate and form weak particle gels at high concentrations (0.5 M NaCl) of salt. Reducing the salt concentration in the emulsions increased the repulsive interactions between the interfacial particles. They found that this minimised the contribution of the particle layer tension to the interfacial energy of the drops [64]. Applying a stress on the order of the Laplace pressure destabilised the emulsions. Whitby and Krebsz [64] observed arrested coalescence for some drop pairs. They also observed that some coalesced drop surfaces were buckled and attributed this to the attached particles making the surfaces of merging drops solid-like in response to compression [64].

5. Harnessing Destabilisation to Fabricate Useful Materials

The developments in our understanding of how interfacial particles affect droplet (and bubble) destabilisation have renewed interest in emulsion phase inversion. This is the key process by which emulsions are transformed into new materials. Phase inversion of Pickering emulsions and foams has produced a wide range of new structures. For example, Binks and Murakami [113] formed powders of particle-stabilised water drops-in-air by inverting the curvature of the air-water surfaces in Pickering foams. Water drops coated with hydrophobic particles roll on solid surfaces like hard spheres [113]. They deform and break open under compressive stress [94]. The powders are used as delivery vehicles for aqueous ingredients in cosmetic products [114].

Coalescence plays a central role in phase inversion. An emulsion of oil drops in water that is being agitated may suddenly transform into an emulsion of water drops in oil in response to a change in the particle wettability, or an increase in the drop volume fraction, as illustrated in Figure 8. The former process is called transitional phase inversion. The latter process is catastrophic. It is achieved by evaporation of the continuous phase, or by pumping the emulsion through narrow spaces. The properties of materials formed by phase inversion are determined by the structural changes that occur as the drops squeeze together and coalesce.

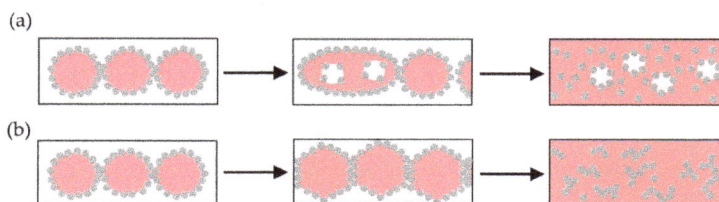

Figure 8. Pickering emulsions destabilise and catastrophically phase invert in response to an increase in their drop volume fraction. (**a**) Inversion may occur via the incorporation of drops of the continuous phase into drops of the dispersed phase as they coalesce in emulsions that are being sheared. Further shearing, or increases of the drop volume fraction, result in inversion of the dispersed and continuous phases in the emulsion; (**b**) Evaporation of the continuous phase from an emulsion can increase the volume fraction of the drops above the maximum fraction at which they can closely pack without distortion. Further evaporation causes the drops to coalesce together into a continuous liquid film and the particles to become completely immersed in the liquid.

New structures are made by manipulating coalescence events in Pickering emulsions at compositions close to phase inversion. For example, Binks and Whitby [115] observed that multiple emulsions form at conditions near catastrophic phase inversion. They proposed that coalescence dominates emulsion formation at compositions where the fraction of the dispersed phase is high [115]. Multiple drops form due to drops of the continuous phase being incorporated into drops of the dispersed phase as they coalesce together (Figure 8a) [115]. The main condition for stabilising multiple emulsions is to stabilise interfaces with both types curvature. Clegg et al. [116] recently reviewed the approaches to making multiple emulsions in the presence of a single type of particle. They argued that the wettability of a fraction of the particle surfaces must be modified by adsorption of air or components from the oil phase.

Monodisperse emulsions can be changed by using a step-wise process to make Pickering emulsions at compositions close to phase inversion. Binks and Rodrigues [117] found that increasing the oil volume fraction in an oil-in-water emulsion at a fixed particle concentration increases the average drop diameter, while reducing the polydispersity of the drop size distribution significantly. This approach forms monodisperse Pickering emulsions by exploiting the limited coalescence that occurs as a large excess of oil-water interface is formed in the presence of a small amount of particles. Direct and inverse emulsions have been fabricated in this way, with r_d ranging from micrometres to millimetres [97,118].

Limited coalescence can also be manipulated to form emulsions at drop volume fractions above the fraction at which drops can be hexagonally close packed without distortion. The highly concentrated emulsions formed are known as high internal phase Pickering emulsions (HIPE). The drop volume fractions in HIPE are so high that they support their own weight and resist mechanical shear. Zang and Clegg [119] investigated the formation of water-in-oil HIPE from mixtures of toluene, water and silanised fumed silica nanoparticles by mechanical mixing. They proposed that HIPE form under conditions where the oil (dispersed) phase is sufficiently viscous to resist being fragmented by shearing and hence inversion [119]. The viscosity of the oil is affected by the particle concentration, shear rate and shearing time. HIPE have been used as templates for making macroporous solids when the liquid phases in the emulsion are removed by evaporation [20,120].

Adelmann et al. [121] reported making organogels by drying dilute oil-in-water Pickering emulsions. The oil phase in the emulsions transformed into a soft, plastic material (an organogel) as the water was removed. Whitby et al. [24] showed that catastrophic phase inversion of o/w Pickering emulsions during evaporation drives the assembly of nanoparticles into network structures within the non-volatile solvent (oil), as depicted in Figure 8b. The elasticity of the oil increases as the volume fraction of the particles in the oil increases. The scaling behavior of the elasticity is consistent with models of the particle networks in the oil as rigid chains connected by permanent, fixed cross-links [24].

Hijnen and Clegg [23] showed that organogels with cellular networks of particles are obtained by evaporating emulsions of partially miscible liquids. They found that closely-packed, polyhedral shapes formed by the drops as they squeezed together were retained in the particle network structure. Their approach utilised the buckling and cracking of particle-laden surfaces observed in emulsions as the two liquid phases mix [57]. Hijnen and Clegg [23] argued that this allowed defects to form in the interfacial layer of particles as the volatile liquid evaporated, but prevented complete rupture. This meant that network structure was not destroyed when the particles became fully wetted by the non-volatile solvent [23].

Oil-in-water Pickering emulsions can be dried to make powders containing encapsulated oil drops. This is achieved by spray drying the emulsions to rapidly remove the water and leave behind powdered solid containing oil. Simovic et al. [122] spray dried mixtures of oil drops (r_d = 200 nm) and silica nanoparticles (r_p = 7 nm) to form micrometre-sized particles of a porous silica matrix encapsulating the oil. Adelmann et al. [121] formed oil powders which did not leak oil for several months by spray drying emulsions with drops (r_d = 10 μm) stabilised by fumed silica nanoparticles (r_p = 30 nm). The powder granules were irregular, fractal-like aggregates, each consisting of several oil drops compressed together. They argued that the particles had sintered together, since the powders could not be re-dispersed in water [121].

The key to producing spray-dried powders that flow freely is controlling the radial distribution of components in the macroscopic droplets of emulsion that are injected into the drier. The structure of the granules formed depends on how the diffusion flux of each component compares to the radial velocity of the shrinking droplet surface [123]. Whitby et al. [124] made powders that could be re-dispersed into particle-coated drops by controlling the number of excess particles present in the emulsions and the extent of flocculation between the excess particles and the particle-stabilised drops.

Bicontinuous materials can be derived from Pickering emulsions undergoing transitional phase inversion. For example, Binks [21] obtained porous solids with bicontinuous-like structures by drying emulsions with compositions close to that required for transitional inversion. Evaporating the oil and water from the emulsions produced solids consisting of interconnected pores that were worm-like in shape. Binks [21] speculated that bicontinuous domains of oil and water formed as the emulsions broke down, since interfaces with both types of curvature could be stabilised by the particles. Direct evidence for the formation of interfaces with zero net curvature in Pickering emulsions at conditions close to inversion was recently obtained. Destribats et al. [125] visualised the interfacial structure in emulsions as the particle wettability was altered systematically. They observed that particles with near neutral wettability show a bimodal distribution of contact angles [125]. The particles formed one of two different contact angles at the oil-water interface, with an average value close to 90°.

Clegg et al. [57] took advantage of this behaviour to stabilise bicontinuous gels (bijels) made from mixtures of partially miscible liquids. The bicontinuous interface formed in a binary liquid as it demixes via spinodal decomposition can be arrested using particles that are neutrally wetted by both liquids. White et al. [126] recently demonstrated that bicontinuous domains form in mixtures of water, 2,6-lutidine and silanised silica nanoparticles (r_p ~400 nm) at compositions near where the mixtures invert from lutidine drop-in-water to water drop-in-lutidine morphologies.

6. Conclusions

Pickering emulsions are metastable colloidal systems. Predicting the lifetime of a Pickering emulsion and how it will break down remains challenging. The stability of an emulsion depends on its composition and the external stresses applied to it during storage and handling. The findings we have reviewed indicate that irreversible changes to Pickering emulsions occur through three mechanisms; particle detachment, mass transfer between the liquid phases and drop coalescence.

Particle detachment from drop surfaces is caused by, for example, adsorbing surfactant on the particle surfaces to amplify their wettability by the external phase. Detachment also occurs under conditions where the particles are strongly flocculated. We suggest that this mechanism will be used as a tool for breaking Pickering emulsions under conditions where the application of physical stress (like heat, shear, magnetic or electrical fields) is undesirable. Akartuna et al. [127], for example, developed a strategy for merging surfactant-stabilised drops on microfluidic chips by chemically inducing pairs of drops to adhere and coalesce.

Mass transfer between the liquid phases in a Pickering emulsion can make the drops shrink and compress together the particles attached to the drop surfaces. The interfacial particles act collectively like a thin, solid shell and buckle under the applied stress. Manipulating the drops in Pickering emulsions into non-spherical shapes in this way is an avenue for fabricating large volumes of non-spherical colloids.

Mass transfer by Ostwald ripening causes larger drops to grow in size. Expansion of particle-laden interfaces may expose pristine drop surfaces and trigger coalescence events in Pickering emulsions. Coalescence is suppressed by adding excess particles to the emulsion. We propose that the ability to disrupt the particle layers coating drops in a transient fashion means it should be possible to fuse together drops of different liquids by destabilising mixtures of Pickering emulsions of different compositions. Fryd and Mason [128] exploited the analogous self-limiting fusion behavior in surfactant-stabilised emulsions to create large volumes of drops with multiple compartments that could encapsulate ingredients of different solubility.

Manipulating drop coalescence is critical to controlling the properties of materials formed by phase inversion. Coalescence between pairs of particle-coated drops due to changes in the surface coverage, the particle wettability or by physical stress has been studied. We think that the next step is to probe coalescence dynamics in bulk emulsions. Feng et al. [129] designed temperature-sensitive surfactants for controlling film formation by a surfactant-stabilised emulsion after visualising how the spatial arrangement of drops influenced phase inversion in the emulsion. Taking a similar approach to characterising how coalescence events propagate in Pickering emulsions will advance the design and performance of products based on the emulsions.

Acknowledgments: Catherine P. Whitby would like to thank Massey University for financial support. Erica J. Wanless would like to thank the Australian Research Council for DP120102305.

Author Contributions: Catherine P. Whitby wrote most of this article. Erica J. Wanless edited the article and contributed Sections 4.3 and 4.5.

Conflicts of Interest: The authors declare no conflict of interest.

References

1. Ghosh, S.; Rousseau, D. Emulsion breakdown in foods and beverages. In *Chemical Deterioration and Physical Instability of Food and Beverages*; Woodhead Publishing: Cambridge, UK, 2010; pp. 260–295.
2. Leal-Calderon, F.; Schmitt, V.; Bibette, J. *Emulsion Science Basic Principles*, 2nd ed.; Springer: New York, NY, USA, 2007.
3. Schramm, L.L. *Emulsions, Foams, and Suspensions: Fundamentals and Applications*; Wiley: Hoboken, NJ, USA, 2005.
4. Tcholakova, S.; Denkov, N.D.; Lips, A. Comparison of solid particles, globular proteins and surfactants as emulsifiers. *Phys. Chem. Chem. Phys.* **2008**, *10*, 1608–1627. [CrossRef] [PubMed]

5. Ramsden, W. Separation of solids in the surface-layers of solutions and 'suspensions' (observations on surface-membranes, bubbles, emulsions, and mechanical coagulation).—Preliminary account. *Proc. R. Soc. Lond.* **1903**, *72*, 156–164. [CrossRef]

6. Pickering, S.U. Emulsions. *J. Chem. Soc. Trans.* **1907**, *91*, 2001–2021. [CrossRef]

7. Aveyard, R.; Binks, B.P.; Clint, J.H. Emulsions stabilised solely by colloidal particles. *Adv. Colloid Interface Sci.* **2003**, *100–102*, 503–546. [CrossRef]

8. Murray, B.S.; Durga, K.; Yusoff, A.; Stoyanov, S.D. Stabilization of foams and emulsions by mixtures of surface active food-grade particles and proteins. *Food Hydrocoll.* **2011**, *25*, 627–638. [CrossRef]

9. He, L.; Lin, F.; Li, X.; Sui, H.; Xu, Z. Interfacial sciences in unconventional petroleum production: From fundamentals to applications. *Chem. Soc. Rev.* **2015**, *44*, 5446–5494. [CrossRef] [PubMed]

10. Sullivan, A.P.; Kilpatrick, P.K. The effects of inorganic solid particles on water and crude oil emulsion stability. *Ind. Eng. Chem. Res.* **2002**, *41*, 3389–3404. [CrossRef]

11. Yan, N.; Masliyah, J.H. Characterization and demulsification of solids-stabilized oil-in-water emulsions part 1. Partitioning of clay particles and preparation of emulsions. *Colloids Surf. A* **1995**, *96*, 229–242. [CrossRef]

12. Crossley, S.; Faria, J.; Shen, M.; Resasco, D.E. Solid nanoparticles that catalyze biofuel upgrade reactions at the water/oil interface. *Science* **2010**, *327*, 68–72. [CrossRef] [PubMed]

13. Leclercq, L.; Mouret, A.; Proust, A.; Schmitt, V.; Bauduin, P.; Aubry, J.-M.; Nardello-Rataj, V. Pickering emulsion stabilized by catalytic polyoxometalate nanoparticles: A new effective medium for oxidation reactions. *Chem. Eur. J.* **2012**, *18*, 14352–14358. [CrossRef] [PubMed]

14. Huang, J.; Yang, H. A ph-switched pickering emulsion catalytic system: High reaction efficiency and facile catalyst recycling. *Chem. Commun.* **2015**, *51*, 7333–7336. [CrossRef] [PubMed]

15. Yang, H.; Zhou, T.; Zhang, W. A strategy for separating and recycling solid catalysts based on the ph-triggered pickering-emulsion inversion. *Angew. Chem. Int. Ed.* **2013**, *52*, 7455–7459. [CrossRef] [PubMed]

16. Van Hee, P.; Hoeben, M.; Van der Lans, R.; Van Der Wielen, L. Strategy for selection of methods for separation of bioparticles from particle mixtures. *Biotech. Bioeng.* **2006**, *94*, 689–709. [CrossRef] [PubMed]

17. Zeng, X.; Osseo-Asare, K. Partitioning behavior of silica in the triton x-100/dextran/water aqueous biphasic system. *J. Colloid Interface Sci.* **2004**, *272*, 298–307. [CrossRef] [PubMed]

18. Binks, B.P.; Clint, J.H.; Fletcher, P.D.I.; Lees, T.J.G.; Taylor, P. Growth of gold nanoparticle films driven by the coalescence of particle-stabilized emulsion drops. *Langmuir* **2006**, *22*, 4100–4103. [CrossRef] [PubMed]

19. Cheng, H.-L.; Velankar, S.S. Film climbing of particle-laden interfaces. *Colloids Surf. A* **2008**, *315*, 275–284. [CrossRef]

20. Barg, S.; Binks, B.; Wang, H.; Koch, D.; Grathwohl, G. Cellular ceramics from emulsified suspensions of mixed particles. *J. Porous Mater.* **2012**, *19*, 859–867. [CrossRef]

21. Binks, B.P. Macroporous silica from solid-stabilized emulsion templates. *Adv. Mater.* **2002**, *14*, 1824–1827. [CrossRef]

22. Ikem, V.O.; Menner, A.; Bismarck, A. High-porosity macroporous polymers sythesized from titania-particle-stabilized medium and high internal phase emulsions. *Langmuir* **2010**, *26*, 8836–8841. [CrossRef] [PubMed]

23. Hijnen, N.; Clegg, P.S. Assembling cellular networks of colloids via emulsions of partially miscible liquids: A compositional approach. *Mater. Horiz.* **2014**, *1*, 360–364. [CrossRef]

24. Whitby, C.P.; Onnink, A.J. Rheological properties and structural correlations in particle-in-oil gels. *Adv. Powder Tech.* **2014**, *25*, 1185–1189. [CrossRef]

25. Denkov, N.D.; Ivanov, I.B.; Kralchevsky, P.A.; Wasan, D.T. A possible mechanism of stabilization of emulsions by solid particles. *J. Colloid Interface Sci.* **1992**, *150*, 589–593. [CrossRef]

26. Tang, J.; Quinlan, P.J.; Tam, K.C. Stimuli-responsive pickering emulsions: Recent advances and potential applications. *Soft Matter* **2015**, *11*, 3512–3529. [CrossRef] [PubMed]

27. Binks, B.P.; Lumsdon, S.O. Influence of particle wettability on the type and stability of surfactant-free emulsions. *Langmuir* **2000**, *16*, 8622–8631. [CrossRef]

28. Levine, S.; Bowen, B.D.; Partridge, S.J. Stabilization of emulsions by fine particles. 1. Partitioning of particles between continuous phase and oil-water interface. *Colloids Surf.* **1989**, *38*, 325–343. [CrossRef]

29. Alargova, R.G.; Warhadpande, D.S.; Paunov, V.N.; Velev, O.D. Foam superstabilization by polymer microrods. *Langmuir* **2004**, *20*, 10371–10374. [CrossRef] [PubMed]

30. Reynaert, S.; Moldenaers, P.; Vermant, J. Control over colloidal aggregation in monolayers of latex particles at the oil−water interface. *Langmuir* **2006**, *22*, 4936–4945. [CrossRef] [PubMed]
31. Subramaniam, A.B.; Mejean, C.; Abkarian, M.; Stone, H.A. Microstructure, morphology, and lifetime of armored bubbles exposed to surfactants. *Langmuir* **2006**, *22*, 5986–5990. [CrossRef] [PubMed]
32. Drelich, A.; Gomez, F.; Clausse, D.; Pezron, I. Evolution of water-in-oil emulsions stabilized with solid particles influence of added emulsifier. *Colloids Surf. A* **2010**, *365*, 171–177. [CrossRef]
33. Katepalli, H.; John, V.T.; Bose, A. The response of carbon black stabilized oil-in-water emulsions to the addition of surfactant solutions. *Langmuir* **2013**, *29*, 6790–6797. [CrossRef] [PubMed]
34. Vashisth, C.; Whitby, C.P.; Fornasiero, D.; Ralston, J. Interfacial displacement of nanoparticles by surfactant molecules in emulsions. *J. Colloid Interface Sci.* **2010**, *349*, 537–543. [CrossRef] [PubMed]
35. Vella, D.; Kim, H.-Y.; Aussillous, P.; Mahadevan, L. Dynamics of surfactant-driven fracture of particle rafts. *Phys. Rev. Lett.* **2006**, *96*, 178301. [CrossRef] [PubMed]
36. Wilde, P.; Mackie, A.; Husband, F.; Gunning, P.; Morris, V. Proteins and emulsifiers at liquid interfaces. *Adv. Colloid Interface Sci.* **2004**, *108*, 63–71. [CrossRef] [PubMed]
37. Mackie, A.R.; Gunning, A.P.; Wilde, P.J.; Morris, V.J. Competitive displacement of β-lactoglobulin from the air/water interface by sodium dodecyl sulfate. *Langmuir* **2000**, *16*, 8176–8181. [CrossRef]
38. Briggs, T.R. Emulsions with finely divided solids. *J. Ind. Eng. Chem.* **1921**, *13*, 1008–1010. [CrossRef]
39. Lucassen-Reynders, E.H.; Tempel, M.V.D. Stabilization of water-in-oil emulsions by solid particles. *J. Phys. Chem.* **1963**, *67*, 731–734. [CrossRef]
40. Horozov, T.S.; Binks, B.P. Particle-stabilized emulsions: A bilayer or a bridging monolayer? *Angew. Chem. Int. Ed.* **2006**, *45*, 773–776. [CrossRef] [PubMed]
41. Lee, M.N.; Chan, H.K.; Mohraz, A. Characteristics of pickering emulsion gels formed by droplet bridging. *Langmuir* **2012**, *28*, 3085–3091. [CrossRef] [PubMed]
42. Xu, H.; Lask, M.; Kirkwood, J.; Fuller, G. Particle bridging between oil and water interfaces. *Langmuir* **2007**, *23*, 4837–4841. [CrossRef] [PubMed]
43. French, D.J.; Taylor, P.; Fowler, J.; Clegg, P.S. Making and breaking bridges in a pickering emulsion. *J. Colloid Interface Sci.* **2015**, *441*, 30–38. [CrossRef] [PubMed]
44. Binks, B.P.; Lumsdon, S.O. Stability of oil-in-water emulsions stabilised by silica particles. *Phys. Chem. Chem. Phys.* **1999**, *1*, 3007–3016. [CrossRef]
45. Horozov, T.S.; Binks, B.P.; Gottschalk-Gaudig, T. Effect of electrolyte in silicone oil-in-water emulsions stabilised by fumed silica particles. *Phys. Chem. Chem. Phys.* **2007**, *9*, 6398–6404. [CrossRef] [PubMed]
46. Chuanuwatanakul, C.; Tallon, C.; Dunstan, D.E.; Franks, G.V. Controlling the microstructure of ceramic particle stabilized foams: Influence of contact angle and particle aggregation. *Soft Matter* **2011**, *7*, 11464–11474. [CrossRef]
47. Whitby, C.P.; Khairul Anwar, H.; Hughes, J. Destabilising pickering emulsions by drop flocculation and adhesion. *J. Colloid Interface Sci.* **2016**, *465*, 158–164. [CrossRef] [PubMed]
48. Ashby, N.P.; Binks, B.P. Pickering emulsions stabilised by laponite clay particles. *Phys. Chem. Chem. Phys.* **2000**, *2*, 5640–5646. [CrossRef]
49. Cates, M.E. Complex Fluids: The Physics of Emulsions. Available online: http://arxiv.org/abs/1209.2290 (accessed on 11 September 2012).
50. Abkarian, M.; Subramaniam, A.B.; Kim, S.-H.; Larsen, R.J.; Yang, S.-M.; Stone, H.A. Dissolution arrest and stability of particle-covered bubbles. *Phys. Rev. Lett.* **2007**, *99*, 188301. [CrossRef] [PubMed]
51. Meinders, M.B.J.; van Vliet, T. The role of interfacial rheological properties on ostwald ripening in emulsions. *Adv. Colloid Interface Sci.* **2004**, *108–109*, 119–126. [CrossRef] [PubMed]
52. Cervantes Martinez, A.; Rio, E.; Delon, G.; Saint-Jalmes, A.; Langevin, D.; Binks, B.P. On the origin of the remarkable stability of aqueous foams stabilised by nanoparticles: Link with microscopic surface properties. *Soft Matter* **2008**, *4*, 1531–1535. [CrossRef]
53. Ettelaie, R.; Murray, B. Effect of particle adsorption rates on the disproportionation process in pickering stabilised bubbles. *J. Chem. Phys.* **2014**, *140*, 204713. [CrossRef] [PubMed]
54. Ettelaie, R.; Murray, B.S. Evolution of bubble size distribution in particle stabilised bubble dispersions: Competition between particle adsorption and dissolution kinetics. *Colloid Surf. A* **2015**, *475*, 27–36. [CrossRef]
55. Avendaño Juárez, J.; Whitby, C.P. Oil-in-water pickering emulsion destabilisation at low particle concentrations. *J. Colloid Interface Sci.* **2012**, *368*, 319–325. [CrossRef] [PubMed]

56. Binks, B.P.; Fletcher, P.D.I.; Holt, B.L.; Kuc, O.; Beaussoubre, P.; Wong, K. Compositional ripening of particle- and surfactant-stabilised emulsions: A comparison. *Phys. Chem. Chem. Phys.* **2010**, *12*, 2219–2226. [CrossRef] [PubMed]

57. Clegg, P.S.; Herzig, E.M.; Schofield, A.B.; Horozov, T.S.; Binks, B.P.; Cates, M.E.; Poon, W.C.K. Colloid-stabilized emulsions: Behaviour as the interfacial tension is reduced. *J. Phys. Condens. Matter* **2005**, *17*, S3433. [CrossRef]

58. Asekomhe, S.O.; Chiang, R.; Masliyah, J.H.; Elliott, J.A.W. Some observations on the contraction behavior of a water-in-oil drop with attached solids. *Ind. Eng. Chem. Res.* **2005**, *44*, 1241–1249. [CrossRef]

59. Datta, S.S.; Shum, H.C.; Weitz, D.A. Controlled buckling and crumpling of nanoparticle-coated droplets. *Langmuir* **2010**, *26*, 18612–18616. [CrossRef] [PubMed]

60. Xu, H.; Melle, S.; Golemanov, K.; Fuller, G. Shape and buckling transitions in solid-stabilized drops. *Langmuir* **2005**, *21*, 10016–10020. [CrossRef] [PubMed]

61. Aveyard, R.; Clint, J.H.; Nees, D.; Paunov, V.N. Compression and structure of monolayers of charged latex particles at air/water and octane/water interfaces. *Langmuir* **2000**, *16*, 1969–1979. [CrossRef]

62. Aveyard, R.; Clint, J.H.; Nees, D.; Quirke, N. Structure and collapse of particle monolayers under lateral pressure at the octane/aqueous surfactant solution interface†. *Langmuir* **2000**, *16*, 8820–8828. [CrossRef]

63. Vella, D.; Aussillous, P.; Mahadevan, L. Elasticity of an interfacial particle raft. *Europhys. Lett.* **2004**, *68*, 212–218. [CrossRef]

64. Whitby, C.P.; Krebsz, M. Coalescence in concentrated pickering emulsions under shear. *Soft Matter* **2014**, *10*, 4848–4854. [CrossRef] [PubMed]

65. Razavi, S.; Cao, K.D.; Lin, B.; Lee, K.Y.C.; Tu, R.S.; Kretzschmar, I. Collapse of particle-laden interfaces under compression: Buckling vs particle expulsion. *Langmuir* **2015**, *31*, 7764–7775. [CrossRef] [PubMed]

66. Aveyard, R.; Binks, B.P.; Clint, J.H.; Fletcher, P.D.I.; Horozov, T.S.; Neumann, B.; Paunov, V.N.; Annesley, J.; Botchway, S.W.; Nees, D.; et al. Measurement of long-range repulsive forces between charged particles at an oil-water interface. *Phys. Rev. Lett.* **2002**, *88*, 246102. [CrossRef] [PubMed]

67. Garbin, V.; Crocker, J.C.; Stebe, K.J. Forced desorption of nanoparticles from an oil–water interface. *Langmuir* **2012**, *28*, 1663–1667. [CrossRef] [PubMed]

68. Arditty, S.; Whitby, C.P.; Binks, B.P.; Schmitt, V.; Leal-Calderon, F. Some general features of limited coalescence in solid-stabilized emulsions. *Eur. Phys. J. E* **2003**, *11*, 273–281. [CrossRef] [PubMed]

69. Wiley, R.M. Limited coalescence of oil droplets in coarse oil-in-water emulsions. *J. Colloid Sci.* **1954**, *9*, 427–437. [CrossRef]

70. Frijters, S.; Günther, F.; Harting, J. Domain and droplet sizes in emulsions stabilized by colloidal particles. *Phys. Rev. E* **2014**, *90*, 042307. [CrossRef] [PubMed]

71. Daware, S.V.; Basavaraj, M.G. Emulsions stabilized by silica rods via arrested demixing. *Langmuir* **2015**, *31*, 6649–6654. [CrossRef] [PubMed]

72. Pauchard, V.; Roy, T. Blockage of coalescence of water droplets in asphaltenes solutions: A jamming perspective. *Colloid Surf. A* **2014**, *443*, 410–417. [CrossRef]

73. Pauchard, V.; Rane, J.P.; Banerjee, S. Asphaltene-laden interfaces form soft glassy layers in contraction experiments: A mechanism for coalescence blocking. *Langmuir* **2014**, *30*, 12795–12803. [CrossRef] [PubMed]

74. Whitby, C.P.; Fornasiero, D.; Ralston, J. Structure of oil-in-water emulsions stabilised by silica and hydrophobised titania particles. *J. Colloid Interface Sci.* **2010**, *342*, 205–209. [CrossRef] [PubMed]

75. Nallamilli, T.; Mani, E.; Basavaraj, M.G. A model for the prediction of droplet size in pickering emulsions stabilized by oppositely charged particles. *Langmuir* **2014**, *30*, 9336–9345. [CrossRef] [PubMed]

76. Zhang, N.; Zhang, L.; Sun, D. Influence of emulsification process on the properties of pickering emulsions stabilized by layered double hydroxide particles. *Langmuir* **2015**, *31*, 4619–4626. [CrossRef] [PubMed]

77. Xu, J.; Ma, A.; Liu, T.; Lu, C.; Wang, D.; Xu, H. Janus-like pickering emulsions and their controllable coalescence. *Chem. Commun.* **2013**, *49*, 10871–10873. [CrossRef] [PubMed]

78. Priest, C.; Reid, M.D.; Whitby, C.P. Formation and stability of nanoparticle-stabilised oil-in-water emulsions in a microfluidic chip. *J. Colloid Interface Sci.* **2011**, *363*, 301–306. [CrossRef] [PubMed]

79. Manga, M.S.; Cayre, O.J.; Williams, R.A.; Biggs, S.; York, D.W. Production of solid-stabilised emulsions through rotational membrane emulsification: Influence of particle adsorption kinetics. *Soft Matter* **2012**, *8*, 1532–1538. [CrossRef]

80. Fan, H.; Striolo, A. Mechanistic study of droplets coalescence in pickering emulsions. *Soft Matter* **2012**, *8*, 9533–9538. [CrossRef]
81. Ata, S. Coalescence of bubbles covered by particles. *Langmuir* **2008**, *24*, 6085–6091. [CrossRef] [PubMed]
82. Ata, S. The detachment of particles from coalescing bubble pairs. *J. Colloid Interface Sci.* **2009**, *338*, 558–565. [CrossRef] [PubMed]
83. Menchaca-Rocha, A.; Martínez-Dávalos, A.; Núñez, R.; Popinet, S.; Zaleski, S. Coalescence of liquid drops by surface tension. *Phys. Rev. E* **2001**, *63*, 046309. [CrossRef] [PubMed]
84. Stover, R.L.; Tobias, C.W.; Denn, M.M. Bubble coalescence dynamics. *AIChE J.* **1997**, *43*, 2385–2392. [CrossRef]
85. Ata, S.; Davis, E.S.; Dupin, D.; Armes, S.P.; Wanless, E.J. Direct observation of ph-induced coalescence of latex-stabilized bubbles using high-speed video imaging. *Langmuir* **2010**, *26*, 7865–7874. [CrossRef] [PubMed]
86. Bournival, G.; Ata, S.; Wanless, E.J. The roles of particles in multiphase processes: Particles on bubble surfaces. *Adv. Colloid Interface Sci.* **2015**, *225*, 114–133. [CrossRef]
87. Tan, S.-Y.; Ata, S.; Wanless, E.J. Direct observation of individual particle armored bubble interaction, stability, and coalescence dynamics. *J. Phys. Chem. B* **2013**, *117*, 8579–8588. [CrossRef] [PubMed]
88. Thompson, K.L.; Giakoumatos, E.C.; Ata, S.; Webber, G.B.; Armes, S.P.; Wanless, E.J. Direct observation of giant pickering emulsion and colloidosome droplet interaction and stability. *Langmuir* **2012**, *28*, 16501–16511. [CrossRef] [PubMed]
89. Taylor, G.I. The formation of emulsions in definable fields of flow. *Proc. R. Soc. Lond. A* **1934**, *146*, 501–523. [CrossRef]
90. Bécu, L.; Benyahia, L. Strain-induced droplet retraction memory in a pickering emulsion. *Langmuir* **2009**, *25*, 6678–6682. [CrossRef] [PubMed]
91. Tan, S.Y.; Tabor, R.F.; Ong, L.; Stevens, G.W.; Dagastine, R.R. Nano-mechanical properties of clay-armoured emulsion droplets. *Soft Matter* **2012**, *8*, 3112–3121. [CrossRef]
92. Ferri, J.K.; Carl, P.; Gorevski, N.; Russell, T.P.; Wang, Q.; Boker, A.; Fery, A. Separating membrane and surface tension contributions in pickering droplet deformation. *Soft Matter* **2008**, *4*, 2259–2266. [CrossRef]
93. Russell, J.T.; Lin, Y.; Böker, A.; Su, L.; Carl, P.; Zettl, H.; He, J.; Sill, K.; Tangirala, R.; Emrick, T.; et al. Self-assembly and cross-linking of bionanoparticles at liquid–liquid interfaces. *Angew. Chemie Int. Ed.* **2005**, *44*, 2420–2426. [CrossRef] [PubMed]
94. Asare-Asher, S.; Connor, J.N.; Sedev, R. Elasticity of liquid marbles. *J. Colloid Interface Sci.* **2015**, *449*, 341–346. [CrossRef] [PubMed]
95. Whitby, C.P.; Fischer, F.E.; Fornasiero, D.; Ralston, J. Shear-induced coalescence of oil-in-water pickering emulsions. *J. Colloid Interface Sci.* **2011**, *361*, 170–177. [CrossRef] [PubMed]
96. Kruglyakov, P.M.; Nushtayeva, A.V.; Vilkova, N.G. Experimental investigation of capillary pressure influence on breaking of emulsions stabilized by solid particles. *J. Colloid Interface Sci.* **2004**, *276*, 465–474. [CrossRef] [PubMed]
97. Arditty, S.; Schmitt, V.; Giermanska-Kahn, J.; Leal-Calderon, F. Materials based on solid-stabilized emulsions. *J. Colloid Interface Sci.* **2004**, *275*, 659–664. [CrossRef] [PubMed]
98. Hermes, M.; Clegg, P.S. Yielding and flow of concentrated pickering emulsions. *Soft Matter* **2013**, *9*, 7568–7575. [CrossRef]
99. Boode, K.; Walstra, P. Partial coalescence in oil-in-water emulsions. 1. Nature of the aggregation. *Colloid Surf. A* **1993**, *81*, 121–137. [CrossRef]
100. Boode, K.; Walstra, P.; Degrootmostert, A.E.A. Partial coalescence in oil-in-water emulsions. 2. Influence of the properties of the fat. *Colloid Surf. A* **1993**, *81*, 139–151. [CrossRef]
101. Fuller, G.T.; Considine, T.; Golding, M.; Matia-Merino, L.; MacGibbon, A. Aggregation behavior of partially crystalline oil-in-water emulsions: Part II—Effect of solid fat content and interfacial film composition on quiescent and shear stability. *Food Hydrocolloids* **2015**, *51*, 23–32. [CrossRef]
102. Arima, S.; Ueno, S.; Ogawa, A.; Sato, K. Scanning microbeam small-angle X-ray diffraction study of interfacial heterogeneous crystallization of fat crystals in oil-in-water emulsion droplets. *Langmuir* **2009**, *25*, 9777–9784. [CrossRef] [PubMed]
103. Rizzo, G.; Norton, J.E.; Norton, I.T. Emulsifier effects on fat crystallisation. *Food Struct.* **2015**, *4*, 27–33. [CrossRef]

104. Thivilliers-Arvis, F.; Laurichesse, E.; Schmitt, V.; Leal-Calderon, F. Shear-induced instabilities in oil-in-water emulsions comprising partially crystallized droplets. *Langmuir* **2010**, *26*, 16782–16790. [CrossRef] [PubMed]

105. Frostad, J.M.; Collins, M.C.; Leal, L.G. Direct measurement of the interaction of model food emulsion droplets adhering by arrested coalescence. *Colloid Surf. A* **2014**, *441*, 459–465. [CrossRef]

106. Pawar, A.B.; Caggioni, M.; Hartel, R.W.; Spicer, P.T. Arrested coalescence of viscoelastic droplets with internal microstructure. *Faraday Disc.* **2012**, *158*, 341–350. [CrossRef]

107. Studart, A.R.; Shum, H.C.; Weitz, D.A. Arrested coalescence of particle-coated droplets into nonspherical supracolloidal structures. *J. Phys. Chem. B* **2009**, *113*, 3914–3919. [CrossRef] [PubMed]

108. Pawar, A.B.; Caggioni, M.; Ergun, R.; Hartel, R.W.; Spicer, P.T. Arrested coalescence in pickering emulsions. *Soft Matter* **2011**, *7*, 7710–7716. [CrossRef]

109. Morse, A.J.; Tan, S.-Y.; Giakoumatos, E.C.; Webber, G.B.; Armes, S.P.; Ata, S.; Wanless, E.J. Arrested coalescence behaviour of giant pickering droplets and colloidosomes stabilised by poly(tert-butylaminoethyl methacrylate) latexes. *Soft Matter* **2014**, *10*, 5669–5681. [CrossRef] [PubMed]

110. Bremond, N.; Thiam, A.R.; Bibette, J. Decompressing emulsion droplets favors coalescence. *Phys. Rev. Lett.* **2008**, *100*, 024501. [CrossRef] [PubMed]

111. Whitby, C.P.; Lotte, L.; Lang, C. Structure of concentrated oil-in-water pickering emulsions. *Soft Matter* **2012**, *8*, 7784–7789. [CrossRef]

112. Tan, J.J.; Zhang, M.; Wang, J.; Xu, J.; Sun, D.J. Temperature induced formation of particle coated non-spherical droplets. *J. Colloid Interface Sci.* **2011**, *359*, 171–178. [CrossRef] [PubMed]

113. Binks, B.P.; Murakami, R. Phase inversion of particle-stabilized materials from foams to dry water. *Nat. Mater.* **2006**, *5*, 865–869. [CrossRef] [PubMed]

114. Yue, S.; Shen, W.; Hapgood, K. Characterisation of liquid marbles in commercial cosmetic products. *Adv. Powder Technol.* **2016**, *27*, 33–41. [CrossRef]

115. Binks, B.P.; Whitby, C.P. Silica particle-stabilized emulsions of silicone oil and water: Aspects of emulsification. *Langmuir* **2004**, *20*, 1130–1137. [CrossRef] [PubMed]

116. Clegg, P.S.; Tavacoli, J.W.; Wilde, P.J. One-step production of multiple emulsions: Microfluidic, polymer-stabilized and particle-stabilized approaches. *Soft Matter* **2016**, *12*, 998–1008. [CrossRef] [PubMed]

117. Binks, B.P.; Rodrigues, J.A. Types of phase inversion of silica particle stabilized emulsions containing triglyceride oil. *Langmuir* **2003**, *19*, 4905–4912. [CrossRef]

118. Arditty, S.; Schmitt, V.; Lequeux, F.; Leal-Calderon, F. Interfacial properties in solid-stabilized emulsions. *Eur. Phys. J. B* **2005**, *44*, 381–393. [CrossRef]

119. Zang, D.; Clegg, P.S. Relationship between high internal-phase pickering emulsions and catastrophic inversion. *Soft Matter* **2013**, *9*, 7042–7048. [CrossRef]

120. Menner, A.; Ikem, V.; Salgueiro, M.; Shaffer, M.S.P.; Bismarck, A. High internal phase emulsion templates solely stabilised by functionalised titania nanoparticles. *Chem. Commun.* **2007**, *41*, 4274–4276. [CrossRef]

121. Adelmann, H.; Binks, B.P.; Mezzenga, R. Oil powders and gels from particle-stabilized emulsions. *Langmuir* **2012**, *28*, 1694–1697. [CrossRef] [PubMed]

122. Simovic, S.; Heard, P.; Hui, H.; Song, Y.; Peddie, F.; Davey, A.K.; Lewis, A.; Rades, T.; Prestidge, C.A. Dry hybrid lipid−silica microcapsules engineered from submicron lipid droplets and nanoparticles as a novel delivery system for poorly soluble drugs. *Mol. Pharm.* **2009**, *6*, 861–872. [CrossRef] [PubMed]

123. Vehring, R.; Foss, W.R.; Lechuga-Ballesteros, D. Particle formation in spray drying. *J. Aerosol. Sci.* **2007**, *38*, 728–746. [CrossRef]

124. Whitby, C.P.; Scarborough, H.; Sian Ng, K.; Ngothai, Y. Spray drying particle-stabilised emulsions. In Proceedings of the Chemeca 2014, Perth, Australia, 28 September–1 October 2014.

125. Destribats, M.; Gineste, S.; Laurichesse, E.; Tanner, H.; Leal-Calderon, F.; Heroguez, V.; Schmitt, V. Pickering emulsions: What are the main parameters determining the emulsion type and interfacial properties? *Langmuir* **2014**, *30*, 9313–9326. [CrossRef] [PubMed]

126. White, K.A.; Schofield, A.B.; Wormald, P.; Tavacoli, J.W.; Binks, B.P.; Clegg, P.S. Inversion of particle-stabilized emulsions of partially miscible liquids by mild drying of modified silica particles. *J. Colloid Interface Sci.* **2011**, *359*, 126–135. [CrossRef] [PubMed]

127. Akartuna, I.; Aubrecht, D.M.; Kodger, T.E.; Weitz, D.A. Chemically induced coalescence in droplet-based microfluidics. *Lab Chip* **2015**, *15*, 1140–1144. [CrossRef] [PubMed]

128. Fryd, M.M.; Mason, T.G. Self-limiting droplet fusion in ionic emulsions. *Soft Matter* **2014**, *10*, 4662–4673. [CrossRef] [PubMed]

129. Feng, H.; Ershov, D.; Krebs, T.; Schroen, K.; Cohen Stuart, M.A.; van der Gucht, J.; Sprakel, J. Manipulating and quantifying temperature-triggered coalescence with microcentrifugation. *Lab Chip* **2015**, *15*, 188–194. [CrossRef] [PubMed]

materials

MDPI

Review

Tuning Amphiphilicity of Particles for Controllable Pickering Emulsion

Zhen Wang and Yapei Wang *

Department of Chemistry, Renmin University of China, Beijing 100872, China; 2012wangzhen@ruc.edu.cn
* Correspondence: yapeiwang@ruc.edu.cn; Tel.: +86-10-6251-9133

Academic Editor: To Ngai
Received: 20 September 2016; Accepted: 4 November 2016; Published: 8 November 2016

Abstract: Pickering emulsions with the use of particles as emulsifiers have been extensively used in scientific research and industrial production due to their edge in biocompatibility and stability compared with traditional emulsions. The control over Pickering emulsion stability and type plays a significant role in these applications. Among the present methods to build controllable Pickering emulsions, tuning the amphiphilicity of particles is comparatively effective and has attracted enormous attention. In this review, we highlight some recent advances in tuning the amphiphilicity of particles for controlling the stability and type of Pickering emulsions. The amphiphilicity of three types of particles including rigid particles, soft particles, and Janus particles are tailored by means of different mechanisms and discussed here in detail. The stabilization-destabilization interconversion and phase inversion of Pickering emulsions have been successfully achieved by changing the surface properties of these particles. This article provides a comprehensive review of controllable Pickering emulsions, which is expected to stimulate inspiration for designing and preparing novel Pickering emulsions, and ultimately directing the preparation of functional materials.

Keywords: Pickering emulsion; particle-stabilized emulsion; particle amphiphilicity; emulsion stability; emulsion phase inversion

1. Introduction

Emulsions have been extensively used in many areas such as cosmetics, the food industry, and material science [1–7]. As a multiphasic mixture system, emulsions typically consist of three main components: oil phase, water phase, and emulsifier [8]. Various emulsifiers, including surfactants [9–12], polymers [13–17], proteins [18–23], and particles [24–27], have been utilized to prepare different kinds of emulsions. Compared with emulsions stabilized by surfactants, polymers, and proteins, particles stabilized emulsions, which are commonly named Pickering emulsions, enjoy characteristic superiority [2]. They make use of micro- or nano-size particles as the interfacial stabilizers, which provide a robust physical barrier against droplet coalescence and retain the long-term stability of emulsions. In addition, Pickering emulsions have low cytotoxicity and good biocompatibility as a result of reducing the use of surfactants [28]. Pickering emulsions as potential candidates are expected to replace traditional emulsions to some extent, and are receiving much attention from industry, bioscience, and materials.

The control of Pickering emulsions plays a pivotal part in many significant processes, for example, oil extraction and recovery [29,30], emulsion polymerization [31–34], and heterogeneous catalysis [35–37]. In this regard, many endeavors have been made in developing controllable Pickering emulsions. Generally, the control of Pickering emulsions involves control over their stability and type which determine the property and performance of the emulsions [8,38]. There are two ways to realize this: changing the compositions of the emulsion phases or tailoring the amphiphilicity of the interfacial emulsifiers. Based on the Ostwald packing theory, it is well known that changing the water-to-oil ratio of emulsion systems can give rise to interconversion between different types of emulsions which is commonly named emulsion phase inversion [8], for example, from water-in-oil (W/O) emulsion to oil-in-water (O/W) emulsion. Together with the water-to-oil ratio, the change of many other parameters such as the concentration of Pickering particles or the components of the two liquid phase, can also induce emulsion phase inversion [39]. These strategies of changing the compositions of emulsion systems can effectively control the type and morphology of Pickering emulsions. However, complicated operation processes and repeated experimental procedures are required to establish the optimum condition for the formation of the desired Pickering emulsions, which cause vast time expenditure and material waste. On the other hand, the addition of excess components including oil phase, water phase, and Pickering particles may damage the original stability of emulsions and lead the emulsions to undergo irreversible changes.

The amphiphilicity of emulsifiers is another crucial factor that notably affects the emulsion stability and type [40–43]. In surfactant stabilized emulsion systems, the amphiphilicity of surfactants is defined as the relative balance between hydrophilic and hydrophobic properties of amphiphiles, which can be well described by the hydrophilic-lipophilic balance (HLB) value [2,8,44]. However, the amphiphilicity in Pickering emulsion systems has different definitions depending on the types of particle. For isotropic particles such as rigid and soft spherical particles, the surface wettability is widely used to describe the amphiphilicity of particles. It can be measured by the three-phase contact angle of particles adsorbed at an oil-water interface [2,44,45]. For anisotropic Janus particles, the concept of Janus balance has been proposed to represent the amphiphilicity of particles in some studies [46,47]. The change of surface wettability or Janus balance can lead to the change of amphiphilicity, which is routinely used to manipulate the stability and type of Pickering emulsions. In contrast to changing the compositions of emulsion systems, tuning amphiphilicity of Pickering particles can direct not only phase inversion between different types of Pickering emulsions, but also reversible emulsification and demulsification via an easier regulating process. This pathway for building controllable Pickering emulsions, as well as their functional materials, is becoming a front-burner issue in the field.

In this review, we highlight some recent advances in tuning the amphiphilicity of particles for preparing controllable Pickering emulsions. Based on the difference of particle species, Pickering emulsifiers are classified into three categories: rigid particles, soft particles, and Janus particles. In the section of rigid particles, we discuss in detail the isotropic rigid particles that are used to control the stability and type of Pickering emulsions. Different mechanisms of tailoring the amphiphilicity of particles are summarized, including molecular adsorption and chemical grafting of small molecules and polymers. For soft particles, examples of isotropic self-assembled objects and microgels are summarized. The next section presents some anisotropic Janus particles that have been successfully used in controlling Pickering emulsions. The amphiphilicity of Janus particles is tuned by changing their surface chemistry and shape. This review is expected to stimulate interest in controlling Pickering emulsion by tuning the amphiphilicity of particles, and to make a contribution in extending the research scope of controllable Pickering emulsions and their future applications in many fields.

2. Rigid Particles

Various unmodified rigid particles have been attempted as emulsifiers to prepare Pickering emulsions in recent years, for example, silica spheres and rods [48–50], metal oxide particles [51,52], clay particles [53], graphene nanosheets [54–56], carbon nanotubes [57,58], carbon black [59], cellulose nanocrystals [60,61], polymeric particles [62,63], lignin particles [64], and chitin particles [65,66]. Stable Pickering emulsions can be successfully formed by using some of these particles. Additionally, the obtained Pickering emulsions have different types depending on the inherent surface wettability of particles. The more hydrophilic particles favor the formation of oil-in-water (O/W) emulsions and the more hydrophobic particles always conduct water-in-oil (W/O) emulsions [2]. However, the particles with specific surface wettability can only form one type of emulsion by keeping the emulsion compositions unchanged. In the absence of stimuli-responsive groups on the surface of particles, it is almost impossible to actualize phase inversion by only using those unmodified particles. For other particles with extremely hydrophilic or hydrophobic surfaces, they generally tend to be dispersed in water or oil phase instead of adsorbing at the oil-water interface, which always causes the emulsification failure or instability of Pickering emulsions. These existing limitations restrict control over Pickering emulsions. In order to easily control the stability and type of Pickering emulsions, tuning the amphiphilicity referring to the controllable change of surface wettability is required for the rigid particles. Among the present strategies for tuning the particle amphiphilicity [28], surface modification is an effective method and has been extensively made use of. According to the interaction mode between particle surface and particular small molecules or polymers, this is generally classified into two categories, including surface adsorption and surface grafting. In this section, we review recent progress related to the two strategies for tuning the amphiphilicity of rigid particles for controllable Pickering emulsions.

2.1. Surface Modification by Non-Covalent Adsorption

2.1.1. Small Molecules

A variety of ingredients have been developed to modify the surface of rigid particles by molecular adsorption of small molecules and polymers based on non-covalent interactions. Among them, surface modification by small molecule adsorption is simple and has attracted great attention from many researchers [53,67,68]. Numerous examples have been presented in which small molecule adsorption was used to tune the amphiphilicity of particles for controlling the stability of Pickering emulsions. For example, Li and coworkers [69] used the adsorption of short-chain aliphatic amines to change the amphiphilicity of Laponite particles, leading to a stable Pickering emulsion. The unmodified Laponite particles are extremely hydrophilic, unable to stabilize the oil-water interface. Therefore, complete phase separation of paraffin oil and water was observed when the raw Laponite particles were used as emulsifiers to prepare Pickering emulsions. Short-chain aliphatic amines including diethylamine (DEA) and trimethylamine (TEA) can adsorb onto the surface of Laponite particles instead of self-assembling into aggregates in solution. Their absorption can effectively increase the hydrophobicity of Laponite particles, facilitating the enrichment of particles at the O/W interface within the emulsion system. The stability of Pickering emulsions could be regulated by controlling the amine concentration. Similar studies have also been demonstrated, for example, methyl orange-modified Laponite particles [70], palmitic acid-modified silica nanoparticles [71], oleic acid-modified silica nanoparticles [72], and magnetite nanoparticles [73], octyl gallate-modified aluminum oxide particles [74] etc. In addition to steady molecules, stimuli-responsive small molecules can transform their molecular structures or conformations under specific external stimuli [9], which enables the smart control of Pickering emulsions. Jiang et al. demonstrated stabilization and destabilization of O/W emulsions by using N′-dodecyl-N,N-dimethylacetamidine modified silica particles [75]. N′-dodecyl-N,N-dimethylacetamidine is a switchable surfactant with a long-chain alkyl group. At high CO_2 concentration, it can be protonated to become positively charged, which can adsorb onto the negatively charged silica particles based on electrostatic interaction. As shown in Figure 1, the molecular adsorption was indicated by the zeta potential of the particles and the adsorption isotherm. The modification of the hydrophobic alkyl group resulted in the wettability change of silica particles from excessive hydrophilicity to partial hydrophobicity, allowing the formation of stable O/W emulsions. Upon purging N_2 or air, the protonated surfactants returned to neutral form and desorbed from the silica particles surface. As a result, the Pickering emulsion was destabilized and phase separation occurred (Figure 1c). Recently, another switchable Pickering emulsion was reported by Zhu and coworkers [76]. As shown in Figure 2, they used a cationic surfactant of cetyltrimethylammonium bromide (CTAB) to change the surface wettability of negatively charged silica particles by electrostatic adsorption. The adsorption of CTAB endowed hydrophilic silica particles with a certain amphiphilicity, which gave a generation of stable O/W emulsions. Subsequently, the obtained Pickering emulsions were destabilized with the addition of an anionic surfactant of sodium dodecyl sulfate (SDS) due to the formation of ion pairs and desorption of CTAB from the particle surface. The stable O/W emulsion could be formed again by adding an equimolar amount of CTAB. The alternate addition of CTAB and SDS in aqueous solution reversibly tuned the amphiphilicity of the silica particles, thus ensuring the control of the destabilization-stabilization behavior of the emulsions.

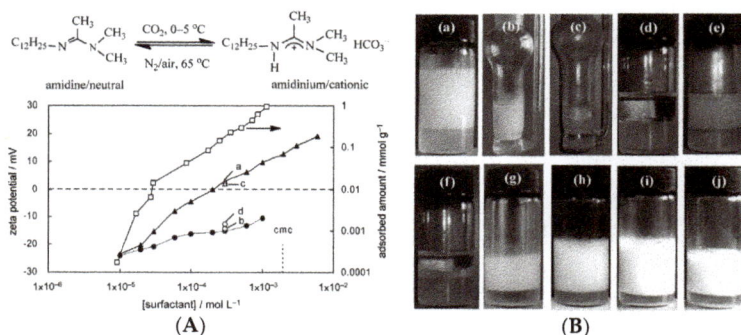

Figure 1. (**A-top**) Schematic illustration of the interconversion between neutral amidine form and cationic amidinium form of N′-dodecyl-N,N-dimethylacetamidine; (**A-bottom**) Zeta potentials of silica nanoparticles that were dispersed in aqueous solutions containing switchable surfactant with cationic amidinium form (▲), and neutral amidine form (●), as a function of initial concentration. The adsorption isotherm of the amidinium form (□) at the silica–water interface is also given as a function of the equilibrium concentration; (**B**) Digital photographs of n-octane-in-water Pickering emulsions that were stabilized by silica nanoparticles and either a switchable surfactant (a–h) undergoing switching or cetyltrimethylammonium bromide (CTAB) (i,j). (**a**) Emulsion with amidinium; (**b**) transfer to a bubbling device; (**c**) bubbling of N_2 through the emulsion; (**d**) transfer to a vial; (**e**) re-homogenization, 24 h later; (**f**) one week later; (**g**) bubbling CO_2 through the emulsion, re-homogenization, one week later; (**h**) emulsion with amidinium after 24 h without bubbling of N_2; (**i**) emulsion with CTAB, 24 h later; (**j**) emulsion with CTAB after bubbling N_2 through the emulsion, 24 h later. Adapted with permission from [75]. Copyright 2013, WILEY-VCH Verlag GmbH & Co. KGaA, Weinheim, Germany.

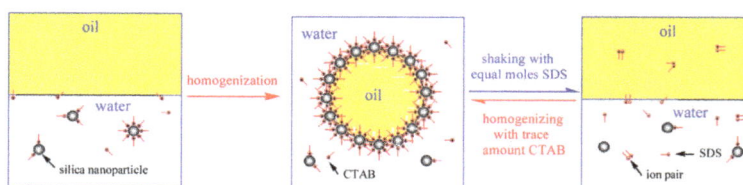

Figure 2. The switch between stable and unstable oil-in-water (O/W) emulsions induced by the alternate addition of a cationic surfactant and an anionic surfactant in the aqueous phase. Reprinted with permission from [76]. Copyright 2015, American Chemical Society.

The phase inversion of Pickering emulsions could be also realized by surface modification of small molecules. Cui and coworkers [77] demonstrated a successful phase inversion from O/W emulsion to W/O emulsion by using a kind of double-chain cationic surfactant-modified silica particle. It was noted that the unmodified silica particles were unable to stabilize the O/W emulsions due to their excessive hydrophilicity. The single-chain cationic surfactants such as dodecyltrimethylammonium bromide (DTAB) and CTAB can adsorb onto the silica particles' surface, therefore increasing the surface hydrophobicity. The adsorption of DTAB or CTAB could improve the stability of O/W emulsions, but the change of particle amphiphilicity was not enough to curve the interface of emulsion into W/O form because of insufficient hydrophobicity of the particles' surface. Compared with single-chain cationic surfactant, a double-chain cationic surfactant of didodecyldimethylammonium bromide (di-C_{12}DMAB) increases the adsorption density of hydrophobic alkyl chains on the particle surface and endows the particle with enhanced hydrophobicity. Their adsorption ultimately induced the emulsion phase inversion from O/W emulsion to W/O emulsion. After that, a similar behavior [78] was observed again by the same group while using $CaCO_3$ nanoparticles and a series of sodium carboxylates with different chain lengths to stabilize Pickering emulsions (Figure 3). The sodium carboxylates have hydrophilic head groups with negative charges and hydrophobic alkyl tails. They can adsorb onto the positively charged $CaCO_3$ nanoparticles' surface based on the combination of electrostatic interaction and hydrophobic effect. With the increase of alkyl chain length, they show enhanced adsorption ability on the particle surface. The adsorbed sodium carboxylates form a monolayer on the $CaCO_3$ nanoparticles with head groups to the particles' surface and hydrophobic tails to water. The arrangement of amphiphiles resulted in the increase of the hydrophobicity of the particles' surface and the change of amphiphilicity. When a low concentration of sodium carboxylate was used, the particle surface reached a certain degree of amphiphilicity, which favored the formation of stable O/W emulsion. As the hydrophobicity was increased to a particular value by increasing the concentration or alkyl chain length of the sodium carboxylates, phase inversion from O/W emulsion to W/O emulsion occurred. Furthermore, for C_{12}Na with a long alkyl chain, bilayer or hemimicelle was formed at higher concentration of sodium carboxylates due to the strong chain-chain interactions, which turned the particles hydrophilic again and caused desorption of particles from the oil-water interface. A second phase inversion from W/O emulsion to O/W emulsion was achieved (Figure 3d).

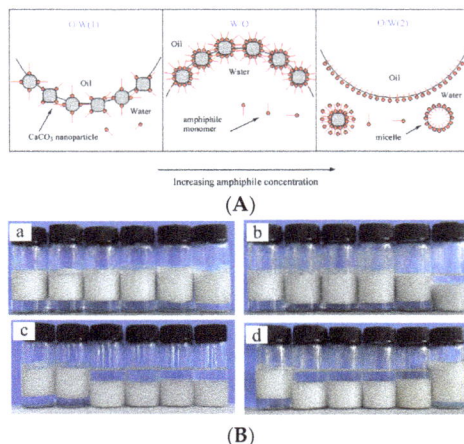

Figure 3. (**A**) Schematic illustration of phase inversion induced by surface adsorption of a series of sodium carboxylates of different chain lengths on $CaCO_3$ nanoparticles; (**B**) Digital photographs of vessels containing toluene-water emulsions stabilized by $CaCO_3$ nanoparticles and sodium carboxylate: (**a**) C_6Na; (**b**) C_8NaI; (**c**) C_{10}Na; and (**d**) C_{12}Na at different concentrations, taken 1 week after preparation. Concentration from left to right: (a–c) 1, 3, 6, 10, 30, 60 mM; (d) 1, 3, 10, 30, 100, 300 mM. Adapted with permission from [78]. Copyright 2012, American Chemical Society.

2.1.2. Polymers

As stated in the above section, surfactants or surfactant-like molecules were usually used for tuning the amphiphilicity of particles. However, those molecules may give rise to toxicity issues in consideration of biological applications and have a potential possibility of uncontrolled fast desorption from particle surfaces [2,40,79]. Compared with small molecules, polymers have more advantages in biocompatibility and adsorption stability. Polymers also possess more varied chain conformations and chain behavior, which can facilitate the formation of complex Pickering emulsions such as double Pickering emulsion [80] and high-internal-phase emulsion (HIPE) [81], and fulfil high-level controllability of Pickering emulsions. Various polymers have been applied to change the particle wettability intending to control the Pickering emulsions. For instance, Wang and coworkers [82] improved the stability of paraffin-water emulsions by using poly(oxypropylene)diamine-modified Laponite particles. Either the Laponite particle or poly(oxypropylene)diamine is a poor emulsifier due to the unfavorable amphiphilicity. When one of them was used alone to prepare the paraffin-water emulsion, complete phase separation into oil and water phase took place. The stable O/W emulsion could be only formed by their combination. It was interpreted that poly(oxypropylene)diamine could adsorb on the particle surface with the two end groups anchored on the surface and the hydrophobic poly(oxypropylene) chain exposed to water, thus changing the amphiphilicity of the particles. Williams et al. [80] used silica particles modified with poly(ethylene imine) (PEI) to prepare a double Pickering emulsion. In the absence of PEI, the hydrophilic silica particles were unable to stabilize any type of Pickering emulsion. Upon absorption of PEI on silica particles with a PEI/silica mass ratio of 0.075, the particles turned partially hydrophobic which enabled the formation of stable O/W emulsions. When the PEI/silica mass ratio was further increased to 0.5, the enhanced surface hydrophobicity led to the phase inversion of emulsion from the O/W form to the W/O form. In short, the interface with different curvature could be formed by using the PEI/silica hybrid particles with a different adsorbed amount of PEI. Based on this mechanism, the double Pickering emulsion of the W/O/W form was also prepared by two-step homogenization (Figure 4). A W/O Pickering emulsion was fabricated at high PEI/silica mass ratio through a first homogenization. This obtained W/O emulsion was subsequently applied to generate the final W/O/W double emulsion at low PEI/silica mass ratio through a second homogenization.

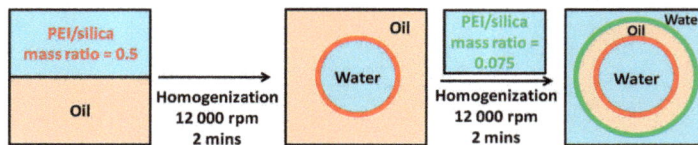

Figure 4. Schematic representation of preparing water-oil-water (W/O/W) double emulsions using poly(ethylene imine)-modified silica particles. Reprinted with permission from [80]. Copyright 2014, American Chemical Society.

Block copolymer is another kind of polymer used to tune the amphiphilicity of particles. Binks and coworkers [83] prepared a polymer-particle complex with tunable amphiphilicity by using the poly[2-(dimethylamino)ethyl methacrylate-block-methyl methacrylate] (PDMA-b-PMMA) to adsorb onto latex particles based on hydrophobic effect. The amphiphilicity of the polymer-particle complexes can be tuned by changing the environmental temperature which has a great influence on the degree of hydration of PDMA chains. At low temperature, the polymer-particle complexes were hydrophilic, which can stabilize an O/W emulsion. With the increase of temperature, the degree of hydration of PDMA chains was reduced. The polymer-particle complexes became more hydrophobic and preferentially wetted by oil. The change in surface wettability induced by increasing temperature brought about phase inversions from O/W emulsions to W/O/W emulsions, finally to W/O emulsions which were formed at the higher temperature. Yoon et al. [84] investigated the effect of poly(acrylic acid) (PAA)-based polymers that were adsorbed onto iron oxide nanoparticles on the morphologies of Pickering emulsions. PAA-based polymers can adsorb onto the iron oxide nanoparticles to change their surface amphiphilicty based on coordination interaction between the carboxylate groups of PAA and the iron. Four PAA-based polymers were attempted to tune the amphiphilicity of iron oxide nanoparticles. The adsorption of the homopolymer PAA could not change the hydrophilicity of iron oxide nanoparticles. Only macroscopic phase separation occurred when these particles acted as emulsifiers during the preparation of Pickering emulsions (Figure 5A). The other three PAA-based polymers are block copolymers poly(acrylic acid-b-butyl acrylate) (PAA-b-PBA) with different PBA block lengths. With the decrease of PBA block length, the interfacial tension of particles at the dodecane/water interface was also decreased. This interfacial change indicated that the modified particles exhibited enhanced interfacial activity. Due to the shortest PBA block length, the particles coated with PAA_{114}-b-PBA_{26} were expected to have the most appropriate amphiphilicity. As shown in Figure 5E, stable emulsions with small droplet size were successfully generated by using these particles. With the increase of PBA block length, the interfacial activity of the particles was lowered, and more nanoparticles were observed to be dispersed in excess aqueous phase. The change of particles in amphiphilicity and decrease of particles' concentration cooperatively led to the increase of emulsion droplet size (Figure 5F,G).

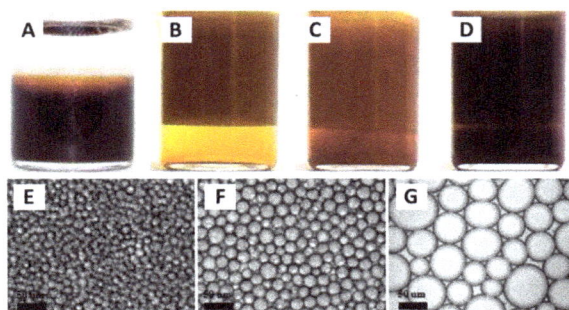

Figure 5. Photographs and microscopy images of oil-in-water emulsions formed between dodecane and aqueous dispersions containing (**A**) poly(acrylic acid) (PAA)-coated NPs; (**B,E**) PAA_{114}-b-PBA_{26}-coated NPs; (**C,F**) PAA_{114}-b-PBA_{38}-coated NPs; and (**D,G**) PAA_{114}-b-PBA_{67}-coated NPs. Images were captured after 1 day at pH = 8 with equal volumes of oil and water phases. Adapted with permission from [84]. Copyright 2012, American Chemical Society.

2.2. Surface Modification by Chemical Grafting

Although molecular adsorption is a simple and effective method to tailor the amphiphilicity of particles, there are still many limitations existing in this method [28]. Molecular adsorption generally has adsorption and desorption equilibrium. This equilibrium crucially relies on the conditions of the system. When the equilibrium conditions are changed, the adsorbed molecules may desorb from the particle surface, which will possibly cause uncontrollable emulsification failure and destabilization of Pickering emulsions. In addition, excessive small molecules or polymers are needed to maintain the equilibrium. The residual molecules, particularly small molecule surfactants or amphiphilic block copolymers, can stabilize the oil-water interface alone. It is difficult to distinguish the exact contributions of the modified particles and the residual molecules to the emulsion stability and phase inversion. The advantages of Pickering emulsion are somehow weakened due to the addition of other amphiphilic molecules. Compared with molecular adsorption, chemical grafting of particular groups on the particle surface is receiving more attention. Chemical grafting requires small molecules or polymers to be fixed on the particle surface by covalent bonds, which ensures less influence against the change of system conditions. Importantly, this strategy can endow particles with various stimuli-responsive groups, for example, temperature, light, pH, CO_2, and ion strength, which reclaims a new vista to control the emulsion type and stability of Pickering emulsions. Similarly, molecules chemically grafted on the particle surface can be small molecules or polymers. In this next section, some recent studies on tuning the amphiphilicity of particles via chemical grafting are reviewed as below.

2.2.1. Small Molecules

Various small molecules have been grafted onto the particle surface to enable the particles to be stimuli-responsive. The capability of responding to external stimuli allows these modified particles to behave as smart emulsifiers for controlling Pickering emulsions. Depending on the properties of small molecules, different external stimuli can be used to actuate the change of particle amphiphilicity. Among them, the amphiphilicity regulation by pH and ion strength has been extensively reported [85–88]. Many reviews have highlighted these excellent works [2,28,45]. It is worth mentioning that tuning the amphiphilicity by CO_2 is becoming a new method. CO_2 is an attractive stimulus owing to its particular advantages of low cost and good biocompatibility [89]. Importantly, it is completely erasable for the emulsion system without any chemical residue contamination. The stability and phase inversion of the emulsion can be readily controlled by only bubbling and releasing CO_2. Liang and coworkers [90] manipulated the stability of Pickering emulsion by using CO_2-responsive particles which were prepared by grafting a CO_2-responsive small molecule of N,N-dimethylacetamide dimethyl acetal (DMA-DMA) on silica particles. The freshly obtained CO_2-responsive particles are able to stabilize an O/W emulsion due to their surface amphiphilicity meeting the basic requirement of forming a stable emulsion. Upon bubbling CO_2 into the emulsion, DMA-DMA was protonated to generate surface charges, and thus the particles became more hydrophilic. This wettability change induced destabilization and phase separation of the emulsion. In contrast, DMA-DMA and phenyl were co-grafted onto silica particles to obtain more hydrophobic particles with completely different initial wettability. Correspondingly, stable W/O emulsions were prepared by these modified particles. When the Pickering emulsion was bubbled with CO_2, a similar phenomenon of emulsion destabilization was observed. In both systems, the stable O/W or W/O emulsions could be recovered by the removal of CO_2 via purging with air (Figure 6).

Figure 6. A schematic illustration for demonstrating the change of CO_2-responsive molecule-modified particle amphiphilicity by bubbling and releasing CO_2 and the resulting stabilization/destabilization of emulsions. Reprinted with permission from [90]. Copyright 2014, American Chemical Society.

Light is another neat and contactless external stimulus [91]. Using light to tune the surface chemistry of particles is considered to be an ideal non-invasive way to control Pickering emulsions, which has been given tremendous attention recently. Chen and coworkers [92] used photochromic spiropyran-grafted up-conversion nanophosphors (Sp-UCNPs) as emulsifiers to establish reversible phase inversion within a Pickering emulsion via light irradiation. In their study, spiropyran was typically grafted on the surface of up-conversion nanophosphors by EDC/NHS chemistry. In terms of conversion of NIR light to UV light by UCNPs, spiropyrans which are responsive to UV light can be triggered from the close-state to the open-state by NIR light. Once photoisomerization occurred, the hydrophobic particles became hydrophilic, leading to the formation of O/W emulsions. Upon the visible light irradiation, spiropyran returned to the close-state so that the particle became hydrophobic again and W/O emulsions were formed. The conversion of emulsion types could be repeated many times by alternating NIR light and visible light irradiation. This novel system was further extended to control biphasic catalysis. Zhang et al. [93] reported a photo-switchable Pickering emulsion system with the use of modified TiO_2 nanoparticles as emulsifiers. Due to the responsive ability of TiO_2 to UV light, reversible regulation of particle amphiphilicity could be accomplished by alternative UV/dark treatment. The raw TiO_2 nanoparticles were grafted with long alkyl chain silanes to change their inherent surface wettability. However, with the amphiphobic property of the particle surface after grafting, the modified TiO_2 nanoparticles were unable to be wetted by oil or water phase and no emulsion was formed. Upon UV irradiation, the excited TiO_2 nanoparticles could induce degradation of the alkyl chain into shorter species. This slight change of surface wettability was enough to retain the stability of the W/O emulsions. With further UV irradiation, alkyl groups were further removed and the particles were covered with hydroxyl groups. The dramatic increase of surface hydrophilicity caused phase inversion and yielded O/W emulsions. Keeping in the dark for a period, the particle surface changed to be hydrophobic and the W/O emulsions were formed again. The different types of emulsions were manipulated by using UV light irradiation and dark-storage alternately (Figure 7).

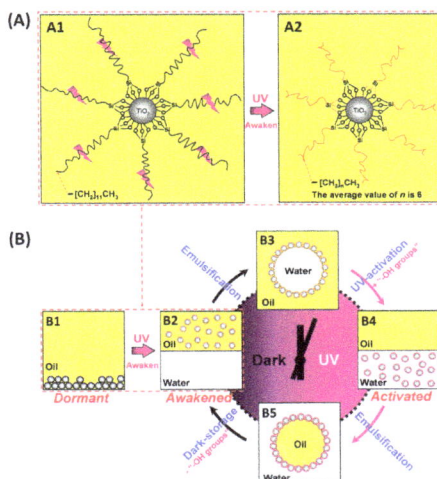

Figure 7. (**A**) Schematic representation of a UV-responsive nanoparticle emulsifier consisting of a TiO_2 core with a shell of long alkyl chain silanes (**A1**) and degraded to short-chain silanes (**A2**) via UV irradiation in hexane; (**B**) Schematic of the switchable Pickering emulsions using TiO_2 nanoparticle emulsifiers under UV/dark treatments: (**B1**) the dormant emulsifiers; (**B2**) the emulsifiers awakened via UV irradiation; (**B3**) the stabilized W/O type emulsions; (**B4**) coalescence and phase separation of the W/O emulsions in response to UV-activation; (**B5**) the stabilized O/W type emulsions and (**B2**) coalescence and phase separation of the W/O emulsions induced by dark storage. The oil is hexane. Adapted with permission from [93]. Copyright 2015, American Chemical Society.

2.2.2. Polymers

Polymers can be also grafted on particles to tailor particle amphiphilicity for controllable Pickering emulsions. Qian and coworkers [94] used a CO_2-responsive poly[2-(diethylamino)ethyl methacrylate] (PDEAEMA) to modify lignin particles through atom transfer radical polymerization (ATRP). The obtained lignin-g-PDEAEMA particles could be dispersed in water in the presence of CO_2 while being flocculated or even precipitated by purging with N_2 quickly (Figure 8). For untreated lignin-g-PDEAEMA particles, decane-in-water emulsions could be stabilized for more than one month. With the addition of CO_2, the O/W emulsion was destabilized into two separated phases. Another two cycles of stabilization and destabilization were also provided by the repeated addition and removal of CO_2 in this study. In addition to CO_2, PDEAEMA is also responsive to temperature. This intriguing property also affords great convenience to switch stabilization and destabilization of Pickering emulsions. In this regard, Tang and coworkers [95] grafted PDMAEMA chains onto the surface of cellulose nanocrystals (CNC) via free radical polymerization. The obtained PDMAEMA-g-CNC particles could adsorb at the oil-water interface and remarkably reduce the interfacial tension, which was beneficial for the formation of stable O/W emulsions. Due to the thermo-responsive character, PDMAEMA chains underwent amphiphilicity change from hydrophilicity to hydrophobicity with the increase of temperature. The O/W emulsions formed at a given pH were destroyed and separated into two phases when they were kept at 50 °C for 5 min (Figure 9). A similar behavior was also observed by the same group using poly(oligoethylene glycol) methacrylate (POEGMA) to modify cellulose nanocrystals [96]. Many other stimuli-responsive polymers have also been reported to tailor the amphiphilicity of particles. However, to our knowledge, only a few light-responsive polymers for controllable Pickering emulsions have been investigated up till now. Enormous opportunities still exist for using non-invasive methods to tune the amphiphilicity of particles by polymer grafting.

Figure 8. Three cycles of the N_2/CO_2-triggered emulsified/demulsified process of lignin-g-PDEAEMA Pickering emulsion. Reprinted with permission from [94]. Copyright 2014, Royal Society of Chemistry.

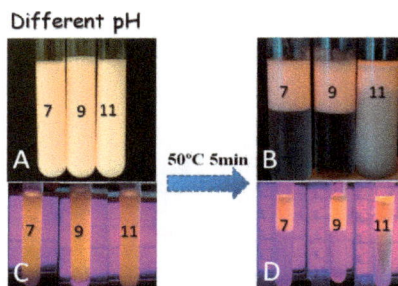

Figure 9. Photographs of toluene-water emulsions stabilized by PDMAEMA-g-CNC at different pH: without UV (**A**); with UV light (**C**); Photographs of emulsions taken after equilibration in a water bath for 5 min (without UV (**B**) and with UV light (**D**)). Reprinted with permission from [95]. Copyright 2014, American Chemical Society.

3. Soft Particles

3.1. Self-Assembled Objects

In addition to the rigid particles as stated above, soft particles have also been applied to prepare Pickering emulsions. Among them, self-assembled objects have shown great promise to stabilize the oil-water interface [97–100] and control the stability of Pickering emulsions [101,102]. Fujii et al. [103] synthesized poly[(ethylene oxide)-block-glycerol monomethacrylate-block-2-(diethylamino)ethyl methacrylate] (PEO-PGMA-PDEA) triblock copolymers, which could be dissolved in aqueous solution to form micelles. After further cross-linking the PGMA blocks using succinic anhydride (SA), the spherical shell cross-linked (SCL) micelles were used as emulsifiers. At high pH value, the micelles possessed particular amphiphilicity as a balance of hydrophobic PDEA core and hydrophilic PEO corona. They were able to stabilize W/O emulsions. At low pH, the hydrophobic PDEA core also became hydrophilic, which significantly influenced the amphiphilicity of micelles. Consequently, micelles detached from the oil-water interface and phase separation of the emulsion happened spontaneously (Figure 10). Ma and coworkers [104] used polyurethane (PU)-based nanoparticles as emulsifiers to control the Pickering emulsions. The nanoparticles were formed by self-assembly of an amphiphilic PU-based grafted copolymer which was synthesized by grafting poly(2-(dimethylamino)ethyl methacrylate) (PDEM) side chains on PU main chain. PDEM is a pH-responsive polymer that is hydrophilic at low pH value and hydrophobic at high pH. As shown in Figure 11, a stable O/W emulsion was formed at a pH range from 3 to 5. Upon changing the pH by the addition of base into the emulsion, PDEM was deprotonated and the particles tended to be wetted by the oil phase. The emulsion was inverted into a W/O type at pH 8–9. With further increase of pH value to 11–12, the particles exhibited hydrophilicity again because of the adsorption of the hydroxyl ions on the particle surfaces. Phase inversion occurred once again and O/W emulsions were formed. Cunningham et al. [105] prepared a kind of nano-object consisting of poly(stearylmethacrylate)–poly(N-2-(methacryloyloxy)ethylpyrrolidone) (PSMA–PNMEP) block copolymer by RAFT dispersion polymerization. Spherical nanoparticles were exclusively obtained with increasing degree of polymerization of PNMEP and keeping the PSMA block length unchanged. Dynamic light scattering (DLS) study indicated nanoparticles from one of these block copolymers PSMA$_{14}$–PNMEP$_{49}$ with an intensity-average diameter of 25 nm. As shown in Figure 12, a hand-shaking W/O emulsion was formed by using these nanoparticles as emulsifiers. However, the emulsion was finally inverted to O/W type after high-speed homogenization. In principle, when the oil and water phases were homogenized under high shear, the nanoparticles were assumed to break up into individual polymer chains which practically stabilized the oil-water interface as polymeric emulsifiers. Combining examinations of DLS, laser diffraction, and transmission electron microscopy, the authors claimed that the PSMA$_{14}$–PNMEP$_{49}$ nanoparticles underwent an inversion from initial hydrophobic spheres to hydrophilic spheres during high-speed homogenization.

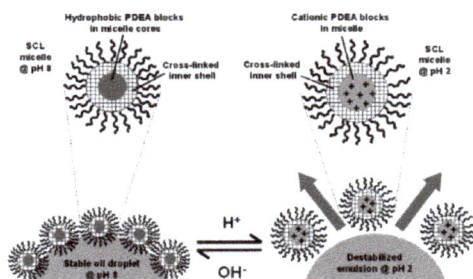

Figure 10. Schematic representation of pH-induced emulsification and demulsification using shell cross-linked micelles as particulate emulsifiers. Dewetting from the oil droplet surface occurs at low pH. Reprinted with permission from [103]. Copyright 2005, American Chemical Society.

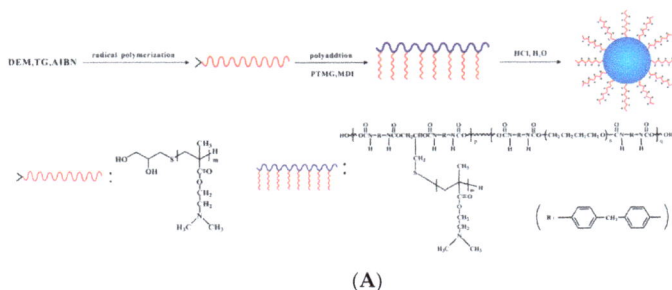

(**A**)

Figure 11. *Cont.*

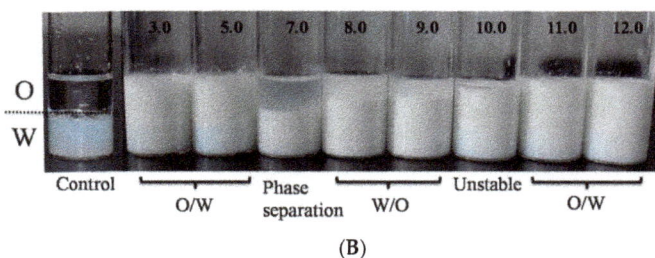

(B)

Figure 11. (**A**) Schematic illustration of the synthesis of PU-g-PDEM graft copolymers and their self-assembly in water; (**B**) The effect of solution pH on the appearance of the emulsion (1:1 water/styrene) after standing for 24 h at 25 °C. Adapted with permission from [104]. Copyright 2013, Royal Society of Chemistry.

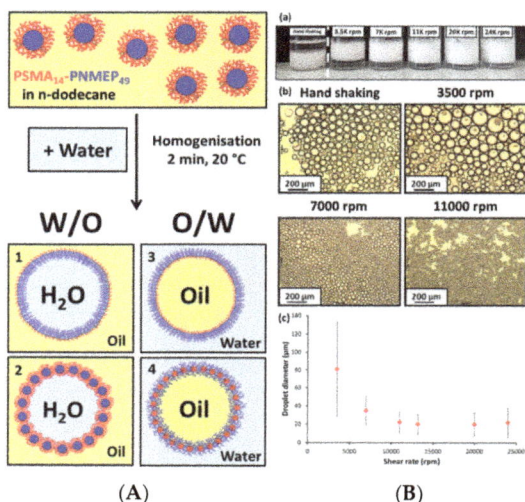

Figure 12. (**A**) Schematic representation of the four possible types of emulsions which could form as a result of homogenizing the poly(stearylmethacrylate)–poly(N-2-(methacryloyloxy)ethylpyrrolidone) (PSMA$_{14}$–PNMEP$_{49}$) nanoparticles prepared in n-dodecane with water; (**B**) (**a**) Digital photographs of Pickering emulsions prepared using PSMA$_{14}$–PNMEP$_{49}$ nanoparticles at various shear rates. Oil-in-water emulsions are formed in all cases, except when hand-shaking is used; this latter approach results in a water-in-oil emulsion instead; (**b**) Optical microscopy images recorded for the droplets prepared via hand-shaking, or via homogenization at 3500 rpm, 7000 rpm or 11,000 rpm (scale bar = 200 μm); (**c**) shear rate dependence for the mean droplet diameter (as determined by laser diffraction) for emulsions prepared using PSMA$_{14}$–PNMEP$_{49}$ spherical nanoparticles as the sole emulsifier. The error bars represent the standard deviation of each mean volume-average droplet diameter, rather than the experimental error. Adapted from [105]. Published by The Royal Society of Chemistry.

3.2. Microgels

Microgels are another kind of soft particles used as emulsifiers to stabilize Pickering emulsions. Many examples involving microgels in emulsion systems have been reported in the past few years [106–112]. Fujii and coworkers [113] synthesized poly(4-vinylpyridine)/silica (P4VP/SiO$_2$) microgels for Pickering emulsion. The P4VP/SiO$_2$ microgel particles were prepared by polymerizing

4-vinylpyridine monomers in the presence of ultrafine aqueous silica sols. The surface of microgel particles consisted of both hydrophilic silica and hydrophobic P4VP chains, which endowed the particle with certain amphiphilicity. When the microgels were involved in the emulsion system, a highly stable emulsion was obtained at high pH value. The particle amphiphilicity was changed if 4-vinylpyridine residues were protonated at low pH. With stepwise decrease of pH value, the degree of protonation was gradually enhanced and the hydrophobic P4VP became hydrophilic. The microgels no longer stabilized the oil-water interface and desorbed from the interface of emulsion, leading to the destabilization and demulsification of the emulsion (Figure 13). After that, this system was further studied by Binks and coworkers [114]. More details about pH-dependent emulsions with the use of P4VP/SiO$_2$ microgels were included. Moreover, the salt effect on the stabilization of Pickering emulsion was also considered in their work. They conclusively demonstrated that the pH value and degree of ionization had significant influences on the stabilization of Pickering emulsions with the use of P4VP/SiO$_2$ microgels as emulsifiers. Ngai et al. [115] investigated another smart emulsion system that is responsive to pH and temperature based on PNIPAM microgels. The PNIPAM microgels were prepared by surfactant-free precipitation copolymerization of N-isopropylacrylamide (NIPAM) and methacrylic acid (MAA). Due to the existence of carboxyl groups, the microgel particles were responsive to pH in addition to temperature due to PNIPAM. Hence, the amphiphilicity of microgel particles could be readily tailored by pH or temperature. As shown in Figure 14, when the microgel particles were used as emulsifiers to prepare the Pickering emulsion, stable emulsions were formed at neutral condition at room temperature. With the decrease of pH or increase of temperature, the hydrophobic component of microgel particles was enhanced, which caused destabilization and phase separation of the emulsions. A similar stimuli-responsive behavior was also observed by Brugger and coworkers using PNIPAM–PMAA microgel particles as emulsifiers and interfacial stabilizers [116,117].

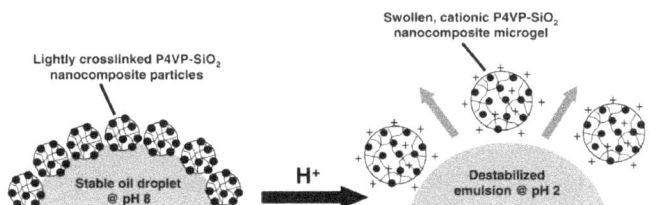

Figure 13. Schematic representation of pH-induced demulsification of an oil-in-water emulsion using lightly cross-linked poly(4-vinylpyridine)/silica (P4VP/SiO$_2$) nanocomposites as particulate emulsifiers. Reprinted with permission from [113]. Copyright 2005, WILEY-VCH Verlag GmbH & Co. KGaA, Weinheim, Germany.

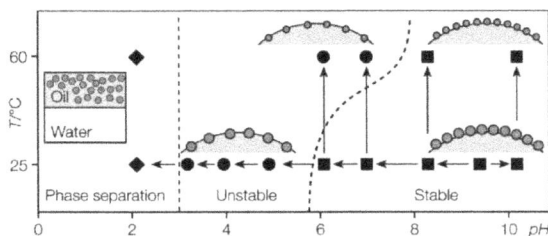

Figure 14. The stabilizing efficiency of poly(N-isopropylacrylamide) (PNIPAM) microgel particles for octanol-in-water emulsions as a function of pH and temperature. ■: Stable, ●: Unstable, ♦: Phase separation. (Arrows indicate the probed transitions). Reprinted with permission from [115]. Copyright 2004, Royal Society of Chemistry.

In addition to the self-assembled objects and microgels, proteins particles have also been exploited to stabilize the Pickering emulsions [118–126]. The emulsion size and stability can be well controlled by changing the pH value or salinity of the systems [127–130].

4. Janus Particles

Janus particles are biphasic colloids consisting of two parts with different properties [25]. Different compositions and shapes of the two parts can endow Janus particles with anisotropy in wetting, optical, electrical, magnetic, and catalytic properties. Compared with homogeneous particles, the anisotropic surface wettability of Janus particles can remarkably lower the interfacial tension and afford stronger interfacial adsorption [131]. With this consideration, a few Pickering emulsion systems were investigated by using various amphiphilic Janus particles as solid emulsifiers in recent years [132–134]. Analogous to the amphiphilicity of molecular amphiphiles such as surfactants and amphiphilic polymers, the amphiphilicity of Janus particles is also important to determine the stability and type of emulsions in these systems. Surface chemistry and shape of the two parts of Janus particles are two important parameters which have been identified as affecting particle amphiphilicity [46,135]. In this section, we highlight some recent examples of controlling Pickering emulsions by tuning the amphiphilicity of Janus particles from these two aspects: changing surface chemistry and particle shape.

4.1. Surface Chemistry

Great efforts have been devoted to tailoring the amphiphilicity of Janus particles by regulating the surface chemistry. The change in amphiphilicity can facilitate the formation of stable emulsions. Xu and coworkers [136] synthesized tadpole-like single chain polymer nanoparticles (TSCPNs) by intramolecularly cross-linking P4VP block of diblock copolymer PMMA$_{2250}$-b-P4VP$_{286}$. As shown in Figure 15A, this Janus particle contains a cross-linked hydrophilic head and a linear hydrophobic tail, which was able to stabilize the W/O emulsion even at a low concentration of 0.0075 wt%. The obtained Pickering emulsion was used as a medium for heterogeneous reaction. In addition, as shown in Figure 15B, this novel Janus particle was also able to stabilize W/O HIPEs as solid emulsifiers at the appropriate conditions [137]. By solidifying the external oil phase and removing the internal water phase of W/O HIPEs, macrocellular polyHIPE materials with interconnected open-cell structures were obtained which were further extended as supporting matrix for loading Pd catalyst. Chen et al. [138] fabricated Janus nanosheets by crushing the polymer-inorganic hybrid hollow spheres which were synthesized by a self-organized sol-gel process forming silica hollow spheres and subsequent polymer grafting onto the interior side. The Janus nanosheets possess a hydrophilic silica layer and a lipophilic polymer layer, ensuring amphiphilicity as well as interfacial activity for emulsion application. The amphiphilic nanosheets were tolerant against solvents and were successfully used as emulsifiers to stabilize W/O emulsion droplets. Fujii et al. [139] partially modified a gold layer on the silica particles by vacuum deposition to prepare Au-SiO$_2$ Janus particles. Due to the hydrophobic property of the Au layer, the surface chemistry of SiO$_2$ particles was partly changed. These Janus particles with amphiphilic performance could stabilize O/W Pickering emulsions for a long time.

The change of surface chemistry can also facilitate the inversion of different emulsion types. Kim and coworkers [135] synthesized a kind of monodisperse bi-compartmentalized Janus microparticle with high throughput by seed monomer swelling and consecutive polymerization. The hydrophilic silica nanoparticles were patched on one of the compartmented bulbs to endow the Janus particles with amphiphilicity. The amphiphilicity of Janus particles was precisely tailored by controlling the relative dimension ratio of the hydrophobic bulb against the whole particle, which was called "degree of Janusity" as the authors stated. When the degree of Janusity was near 0.25, W/O emulsions were formed. As shown in Figure 16D, with the increase of the degree of Janusity, the change in amphiphilicity enabled the phase inversion from W/O emulsion to O/W emulsion and the stability of the generated emulsion was also enhanced. Zhao et al. [140] grafted the temperature responsive

PNIPAM and pH responsive PDEAEMA on two sides of a silica nanosheet to prepare a dually responsive Janus particle. The amphiphilicity of Janus particles could be tailored by changing the pH or environmental temperature. Different types of emulsions were formed and transformed each other by using these Janus particles with tunable amphiphilicity. At high pH and high temperature, Janus particles were completely hydrophobic, which were only dispersed in the toluene phase. At low pH and low temperature, the Janus particles on the contrary were hydrophilic and dispersed in the water phase. With proper change of pH and temperature, the Janus particles showed amphiphilicity and could stabilize the oil-water interface. At low pH and high temperature, O/W emulsions were formed. At high pH and low temperature, W/O emulsions were preferred. The control of phase inversion of emulsions was achieved by a combination of pH and temperature (Figure 17).

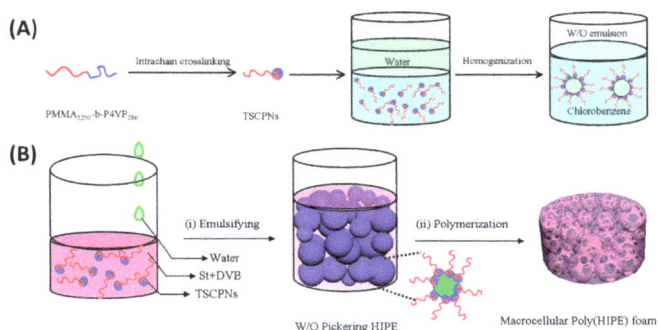

Figure 15. (**A**) Water in Chlorobenzene W/O Emulsions Stabilized by tadpole-like single chain polymer nanoparticles (TSCPNs). Reprinted with permission from [136]. Copyright 2014, American Chemical Society; (**B**) Schematic representation of macrocellular polyHIPE foam templated from TSCPNs stabilized W/O HIPE template. Reprinted with permission from [137]. Copyright 2015, Royal Society of Chemistry.

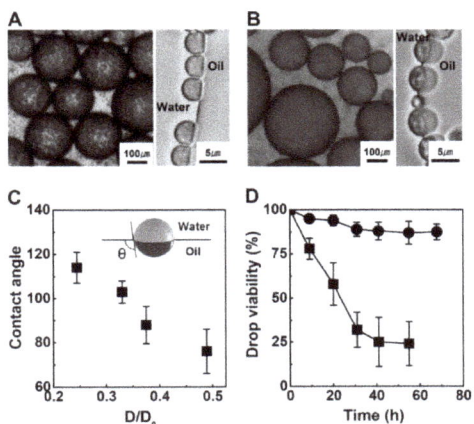

Figure 16. Bright-field microscopy images of the Pickering emulsions stabilized by silica-NP-patched Janus particles with the different degrees of Janusity: (**A**) $D/D_0 = 0.25$; and (**B**) $D/D_0 = 0.5$. The inset images show adhesion of amphiphilic Janus particles at the hexadecane–water interface; (**C**) Contact angles of silica NP-patched Janus particles at the hexadecane–water interface; (**D**) Viability of Pickering emulsion drops at 50 °C: $D/D_0 = 0.25$ (■) and $D/D_0 = 0.5$ (●). Reprinted with permission from [135]. Copyright 2016, WILEY-VCH Verlag GmbH & Co. KGaA, Weinheim, Germany.

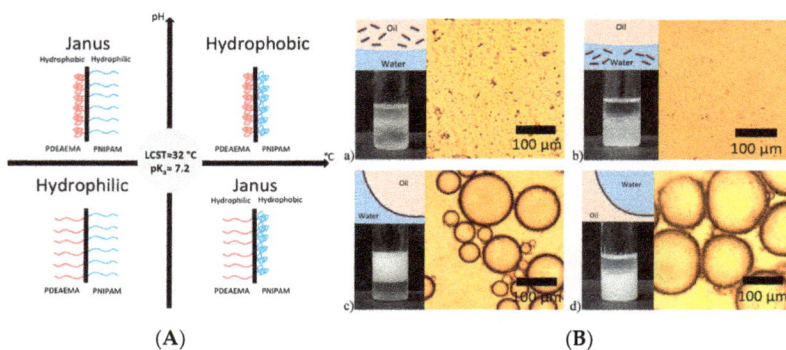

Figure 17. (**A**) Schematic illustration of pH and temperature dually responsive Janus composite nanosheets; (**B**) Emulsification of the immiscible mixture of toluene/water with the PNIPAM/silica/ PDEAEMA Janus composite nanosheets. (**a**) Optical microscopy image of the mixture (inset) at pH = 10 and T = 50 °C, no emulsification occurs, the hydrophobic nanosheets dispersible in toluene; (**b**) at pH = 2 and T = 25 °C, no emulsification occurs, the hydrophilic nanosheets dispersible in water; (**c**) a toluene-in-water emulsion forms at pH = 2 and T = 50 °C; (**d**) a water-in-toluene emulsion forms at pH = 10 and T = 25 °C. Adapted with permission from [140]. Copyright 2015, American Chemical Society.

4.2. Shape

The shape of the Janus particles is another factor that has a significant influence on their interfacial behavior. Ruhland and coworkers [141] investigated in detail the interfacial behavior of Janus particles with different shapes including Janus spheres, Janus cylinders, and Janus discs by a combination of dynamic interfacial tension measurements and computer simulations. The different adsorption kinetics and equilibrium values of the interfacial tension with the use of different Janus particles were compared (Figure 18). The particle shape is considered to affect the surface activity of Janus particles. The stability and type of Pickering emulsion stabilized by these Janus particles show close dependence on the shape of the Janus particles. Multiple shaped Janus particles have been practically exploited as solid emulsifiers for the formation of stable emulsions and controllable inversion of different emulsion types. A snowman-like Janus particle consisting of a hydrophilic bulb and a hydrophobic bulb was synthesized based on the seeded polymerization technique by Kim and coworkers [142]. Stable emulsions of spheres, ellipsoids, and cylinders could be formed due to the substantial absorption of these amphiphilic Janus particles at the O/W interface. The inversion of different emulsion types was also achieved by changing the relative ratio of hydrophilic and hydrophobic bulbs in the snowman-like particles by Liu and coworkers [46]. Different from the seed polymerization technique, the snowman-like Janus particles in their study were prepared by extruding a lobe based on swelling the polymeric core from a spherical core-shell structure. By tuning the monomer/particle weight ratio, the relative size ratio of the hydrophilic inorganic part and the hydrophobic polymeric lobe could be tailored. When the Janus particles with a large hydrophobic polymeric lobe were used to stabilize the Pickering emulsion, a W/O emulsion was formed. With the decrease of hydrophobic lobe size, the emulsion was inverted into an O/W type (Figure 19).

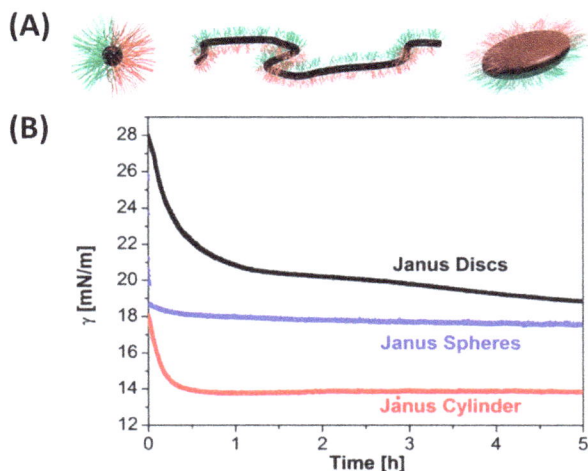

Figure 18. (**A**) Overview of possible Janus particle architectures: Spheres, Cylinders, and Discs; (**B**) Influence of the Janus particle shape on the interfacial tension. Interfacial tension isotherms of solutions of Janus particles in toluene at a water/toluene interface. Adapted with permission from [141]. Copyright 2013, American Chemical Society.

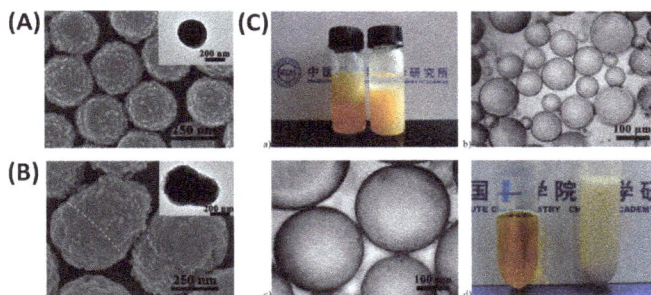

Figure 19. Morphological evolution of the anisotropic composite particles at varied monomer/particle weight ratio: (**A**) 2:1; (**B**) 16:1; (**C**) Janus characteristics of the anisotropic particles. (**a**) Left: a decane-in-water emulsion stabilized with the Janus particle as shown in Figure (**A**); right: a water-in-decane emulsion stabilized with the Janus particle as shown in Figure (**B**); water/decane volume ratio is 2:1, and the particle/water weight ratio is 1:100; (**b**) Optical microscopy images of the decane-in-water emulsion; and (**c**) water-in-decane emulsion; (**d**) Left: immiscible toluene/water mixture; right: a water-in-toluene emulsion stabilized with the Janus particle as shown in Figure 5d. Water/toluene volume ratio is 2:1, and the particle/water weight ratio is 2:1000. Adapted with permission from [46]. Copyright 2012, American Chemical Society.

Another novel "mushroom-like" Janus particle [143] was reported to control the stability of Pickering emulsions by Yamagami and coworkers (Figure 20). The poly(methyl methacrylate)/poly(styrene-2-(2-bromoisobutyryloxy)ethyl methacrylate)-graft-poly(2-(dimethyl amino)ethyl methacrylate) PMMA/P(S-BIEM)-g-PDM Janus particles were synthesized by using a surface-initiated activator generated by electron transfer (AGET) ATRP to graft PDM on the P(S-BIEM) side of the composite particles based on phase separation under solvent evaporation. The shape anisotropy of the Janus particles could be precisely tailored by changing the size of the P(S-BIEM) side, which was conducted by controlling the

P(S-BIEM) content in the composite particles. The introduction of PDM enables the Janus particles to be dually responsive to pH and temperature. Janus particles with controllable shapes were used as solid emulsifiers to stabilize 1-octanol-in-water emulsion. The stability of the emulsion could be controlled by altering the temperature and pH. Tu et al. [144] fabricated a shape switchable Janus particle to control the phase inversion of Pickering emulsions. The Janus particles were synthesized by seeded emulsion polymerization and subsequent acid hydrolysis. Due to a pH-responsive ability, the particle part with acrylic acid blocks were swollen at high pH and deswollen at low pH, thus making the particle shape controllable by pH. As shown in Figure 21, the Janus particles remained oblate-like or almost spherical shape at pH 2.2 or in a deionized water medium, which favored the formation of W/O emulsions. With increasing pH to 11.0, the shape of Janus particles was transformed into a dumbbell. The interfacial activity of the Janus particles also changed with the shape change, inducing the emulsions with different types. W/O and O/W emulsions could be reversibly switched by adding a small amount of acid or base in the emulsion system. This study was claimed as the first example of transitional phase inversion of Pickering emulsions by dynamically tuning the shape to reversibly change the amphiphilicity of the Janus particles.

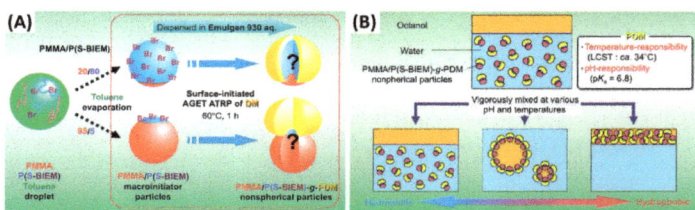

Figure 20. (**A**) Preparation of mushroom-like Janus polymer particles by site-selective surface-initiated activator generated by electron transfer (AGET) ATRP in aqueous dispersed systems; (**B**) The scheme of whole behaviors of Pickering emulsion formed by stimuli-responsive "mushroom-like" Janus polymer particles. Adapted with permission from [143]. Copyright 2014, American Chemical Society.

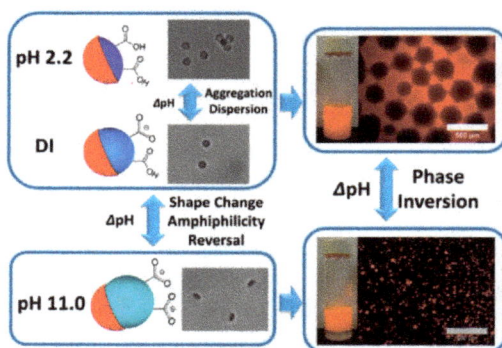

Figure 21. Janus particles change their aggregation/dispersion behavior and also transform into different shapes in response to pH changes. Janus particles with tunable amphiphilicity can stabilize different types of emulsions (oil-in-water and water-in-oil). Reprinted with permission from [144]. Copyright 2014, American Chemical Society.

5. Conclusions

In this article, we comprehensively reviewed recent progress in tuning the amphiphilicity of particles for controllable Pickering emulsions. The whole map of amphiphilicity regulation was presented on the basis of particle species. Rigid particles can be used to control the stability and type of Pickering emulsion by molecular adsorption or chemical grafting of functional substances. Soft particles including self-assembled objects and microgels were also highlighted for the control of stabilization and destabilization of emulsions. Anisotropic Janus particles most likely bridging two immiscible phases together were highly emphasized. Their amphiphilicity, referring to the interfacial activity, was tuned by changing the particle surface chemistry or shape. From the aspect of amphiphilicity, this review gives new bearing on the preparation of controllable emulsions, though most examples did not correlate the controllable Pickering emulsions with the amphiphilicity of the particles. It is believed that this concept will open a new avenue to guide the preparation of functional materials based on emulsion technique. However, there are still existing opportunities and challenges in this area. Novel particles are expected to be synthesized for the preparation of complex Pickering emulsions including double emulsions and HIPEs, which requires the particles to stabilize at least two oil-water interfaces at the same time. More advanced emulsions such as W/W or O/O emulsions are hoped to be exploited by nanoparticles with ideal amphiphilicity. In addition, Pickering emulsion systems as a result of tuning amphiphilicity may be manipulated by biological actuators, which will meliorate the toxicity issue in enzymatic and clinical applications.

Acknowledgments: This review was financially supported by the National Natural Science Foundation of China (51373197, 21422407, and 21674127).

Author Contributions: Z.W. retrieved the literature involved in this review and drafted the manuscript. Y.W. commanded the design of this review and revised the article.

Conflicts of Interest: The authors declare no conflict of interest.

References

1. Wang, W.; Zhang, M.; Chu, L. Functional polymeric microparticles engineered from controllable microfluidic emulsions. *Acc. Chem. Res.* **2014**, *47*, 373–384. [CrossRef] [PubMed]
2. Tang, J.; Quinlan, P.J.; Tam, K.C. Stimuli-responsive Pickering emulsions: Recent advances and potential applications. *Soft Matter* **2015**, *11*, 3512–3529. [CrossRef] [PubMed]
3. Huang, X.; Qian, Q.; Zhang, X.; Du, W.; Xu, H.; Wang, Y. Assembly of carbon nanotubes on polymer particles: Towards rapid shape change by near-infrared light. *Part. Part. Syst. Charact.* **2013**, *30*, 235–240. [CrossRef]
4. Qian, Q.; Huang, X.; Zhang, X.; Xie, Z.; Wang, Y. One-step preparation of macroporous polymer particles with multiple interconnected chambers: A candidate for trapping biomacromolecules. *Angew. Chem. Int. Ed.* **2013**, *52*, 10625–10629. [CrossRef] [PubMed]
5. Wang, J.; Zhao, J.; Li, Y.; Yang, M.; Chang, Y.; Zhang, J.; Sun, Z.; Wang, Y. Enhanced light absorption in porous particles for ultra-NIR-sensitive biomaterials. *ACS Macro Lett.* **2015**, *4*, 392–397. [CrossRef]
6. Cao, Y.; Wang, Z.; Liao, S.; Wang, J.; Wang, Y. A light-activated microheater for the remote control of enzymatic catalysis. *Chem. Eur. J.* **2016**, *22*, 1152–1158. [CrossRef] [PubMed]
7. Liao, S.; He, Y.; Wang, D.; Dong, L.; Du, W.; Wang, Y. Dynamic interfacial printing for monodisperse droplets and polymeric microparticles. *Adv. Mater. Technol.* **2016**, *1*, 1600021. [CrossRef]
8. Kumar, A.; Li, S.; Cheng, C.-M.; Lee, D. Recent developments in phase inversion emulsification. *Ind. Eng. Chem. Res.* **2015**, *54*, 8375–8396. [CrossRef]
9. Brown, P.; Butts, C.P.; Eastoe, J. Stimuli-responsive surfactants. *Soft Matter* **2013**, *9*, 2365–2374. [CrossRef]
10. Liu, Y.; Jessop, P.G.; Cunningham, M.; Eckert, C.A.; Liotta, C.L. Switchable surfactants. *Science* **2006**, *313*, 958–960. [CrossRef] [PubMed]
11. Pradhan, M.; Rousseau, D. A one-step process for oil-in-water-in-oil double emulsion formation using a single surfactant. *J. Colloid Interface Sci.* **2012**, *386*, 398–404. [CrossRef] [PubMed]
12. Huang, X.; Qian, Q.; Wang, Y. Anisotropic particles from a one-pot double emulsion induced by partial wetting and their triggered release. *Small* **2014**, *10*, 1412–1420. [CrossRef] [PubMed]
13. Macon, A.L.B.; Rehman, S.U.; Bell, R.V.; Weaver, J.V.M. Reversible assembly of pH responsive branched copolymer-stabilised emulsion via electrostatic forces. *Chem. Commun.* **2016**, *52*, 136–139. [CrossRef] [PubMed]

14. Ku, K.H.; Shin, J.M.; Klinger, D.; Jang, S.G.; Hayward, R.C.; Hawker, C.J.; Kim, B.J. Particles with tunable porosity and morphology by controlling interfacial instability in block copolymer emulsions. *ACS Nano* **2016**, *10*, 5243–5251. [CrossRef] [PubMed]

15. Hanson, J.A.; Chang, C.B.; Graves, S.M.; Li, Z.B.; Mason, T.G.; Deming, T.J. Nanoscale double emulsions stabilized by single-component block copolypeptides. *Nature* **2008**, *455*, 85–88. [CrossRef] [PubMed]

16. Hong, L.Z.; Sun, G.Q.; Cai, J.G.; Ngai, T. One-step formation of w/o/w multiple emulsions stabilized by single amphiphilic block copolymers. *Langmuir* **2012**, *28*, 2332–2336. [CrossRef] [PubMed]

17. Wang, D.; Xiao, L.; Zhang, X.; Zhang, K.; Wang, Y. Emulsion templating cyclic polymers as microscopic particles with tunable porous morphology. *Langmuir* **2016**, *32*, 1460–1467. [CrossRef] [PubMed]

18. Patel, A.R.; Velikov, K.P. Zein as a source of functional colloidal nano- and microstructures. *Curr. Opin. Colloid Interface Sci.* **2014**, *19*, 450–458. [CrossRef]

19. Filippidi, E.; Patel, A.R.; Bouwens, E.C.M.; Voudouris, P.; Velikov, K.P. All-natural oil-filled microcapsules from water-insoluble proteins. *Adv. Funct. Mater.* **2014**, *24*, 5962–5968. [CrossRef]

20. McClements, D.J. Protein-stabilized emulsions. *Curr. Opin. Colloid Interface Sci.* **2004**, *9*, 305–313. [CrossRef]

21. Wilde, P.; Mackie, A.; Husband, F.; Gunning, P.; Morris, V. Proteins and emulsifiers at liquid interfaces. *Adv. Colloid Interface* **2004**, *108–109*, 63–71. [CrossRef] [PubMed]

22. Damodaran, S. Protein stabilization of emulsions and foams. *J. Food Sci.* **2005**, *70*, R54–R66. [CrossRef]

23. Dickinson, E. Flocculation of protein-stabilized oil-in-water emulsions. *Colloids Surf. B Biointerfaces* **2010**, *81*, 130–140. [CrossRef] [PubMed]

24. Boker, A.; He, J.; Emrick, T.; Russell, T.P. Self-assembly of nanoparticles at interfaces. *Soft Matter* **2007**, *3*, 1231–1248. [CrossRef]

25. Kumar, A.; Park, B.J.; Tu, F.Q.; Lee, D. Amphiphilic Janus particles at fluid interfaces. *Soft Matter* **2013**, *9*, 6604–6617. [CrossRef]

26. Tavernier, I.; Wijaya, W.; Van der Meeren, P.; Dewettinck, K.; Patel, A.R. Food-grade particles for emulsion stabilization. *Trends Food Sci. Technol.* **2016**, *50*, 159–174. [CrossRef]

27. Lam, S.; Velikov, K.P.; Velev, O.D. Pickering stabilization of foams and emulsions with particles of biological origin. *Curr. Opin. Colloid Interface Sci.* **2014**, *19*, 490–500. [CrossRef]

28. Chevalier, Y.; Bolzinger, M.A. Emulsions stabilized with solid nanoparticles: Pickering emulsions. *Colloids Surf. A* **2013**, *439*, 23–34. [CrossRef]

29. Wang, X.F.; Shi, Y.; Graff, R.W.; Lee, D.; Gao, H.F. Developing recyclable pH-responsive magnetic nanoparticles for oil-water separation. *Polymer* **2015**, *72*, 361–367. [CrossRef]

30. Chen, Y.; Bai, Y.; Chen, S.; Jup, J.; Li, Y.; Wang, T.; Wang, Q. Stimuli-responsive composite particles as solid-stabilizers for effective oil harvesting. *ACS Appl. Mater. Interfaces* **2014**, *6*, 13334–13338. [CrossRef] [PubMed]

31. Walther, A.; Hoffmann, M.; Mueller, A.H.E. Emulsion polymerization using Janus particles as stabilizers. *Angew. Chem. Int. Ed.* **2008**, *47*, 711–714. [CrossRef] [PubMed]

32. Voorn, D.J.; Ming, W.; van Herk, A.M. Polymer-clay nanocomposite latex particles by inverse Pickering emulsion polymerization stabilized with hydrophobic montmorillonite platelets. *Macromolecules* **2006**, *39*, 2137–2143. [CrossRef]

33. Tuncer, C.; Samav, Y.; Ulker, D.; Baker, S.B.; Butun, V. Multi-responsive microgel of a water-soluble monomer via emulsion polymerization. *J. Appl. Polym. Sci.* **2015**, *132*. [CrossRef]

34. Haaj, S.B.; Thielemans, W.; Magnin, A.; Boufi, S. Starch nanocrystal stabilized Pickering emulsion polymerization for nanocomposites with improved performance. *ACS Appl. Mater. Interfaces* **2014**, *6*, 8263–8273. [CrossRef] [PubMed]

35. Pera-Titus, M.; Leclercq, L.; Clacens, J.-M.; De Campo, F.; Nardello-Rataj, V. Pickering interfacial catalysis for biphasic systems: From emulsion design to green reactions. *Angew. Chem. Int. Ed.* **2015**, *54*, 2006–2021. [CrossRef] [PubMed]

36. Chen, Z.; Zhao, C.; Ju, E.; Ji, H.; Ren, J.; Binks, B.P.; Qu, X. Design of surface-active artificial enzyme particles to stabilize Pickering emulsions for high-performance biphasic biocatalysis. *Adv. Mater.* **2016**, *28*, 1682–1688. [CrossRef] [PubMed]

37. Huang, J.; Cheng, F.; Binks, B.P.; Yang, H. pH-responsive gas-water-solid interface for multiphase catalysis. *J. Am. Chem. Soc.* **2015**, *137*, 15015–15025. [CrossRef] [PubMed]

38. Whitby, C.P.; Wanless, E.J. Controlling Pickering emulsion destabilisation: A route to fabricating new materials by phase inversion. *Materials* **2016**, *9*, 626. [CrossRef]

39. Tcholakova, S.; Denkov, N.D.; Lips, A. Comparison of solid particles, globular proteins and surfactants as emulsifiers. *Phys. Chem. Chem. Phys.* **2008**, *10*, 1608–1627. [CrossRef] [PubMed]

40. Huang, X.; Fang, R.; Wang, D.; Wang, J.; Xu, H.; Wang, Y.; Zhang, X. Tuning polymeric amphiphilicity via Se-N interactions: Towards one-step double emulsion for highly selective enzyme mimics. *Small* **2015**, *11*, 1537–1541. [CrossRef] [PubMed]

41. Huang, X.; Yang, Y.; Shi, J.; Ngo, H.T.; Shen, C.; Du, W.; Wang, Y. High-internal-phase emulsion tailoring polymer amphiphilicity towards an efficient NIR-sensitive bacteria filter. *Small* **2015**, *11*, 4876–4883. [CrossRef] [PubMed]

42. Chen, Y.; Wang, Z.; Wang, D.; Ma, N.; Li, C.; Wang, Y. Surfactant-free emulsions with erasable triggered phase inversions. *Langmuir* **2016**, *32*, 11039–11042. [CrossRef] [PubMed]

43. Wang, D.; Huang, X.; Wang, Y. Managing the phase separation in double emulsion by tuning amphiphilicity via a supramolecular route. *Langmuir* **2014**, *30*, 14460–14468. [CrossRef] [PubMed]

44. Binks, B.P. Particles as surfactants - similarities and differences. *Curr. Opin. Colloid Interface Sci.* **2002**, *7*, 21–41. [CrossRef]

45. Leal-Calderon, F.; Schmitt, V. Solid-stabilized emulsions. *Curr. Opin. Colloid Interface Sci.* **2008**, *13*, 217–227. [CrossRef]

46. Liu, B.; Liu, J.; Liang, F.; Wang, Q.; Zhang, C.; Qu, X.; Li, J.; Qiu, D.; Yang, Z. Robust anisotropic composite particles with tunable Janus balance. *Macromolecules* **2012**, *45*, 5176–5184. [CrossRef]

47. Jiang, S.; Granick, S. Controlling the geometry (Janus balance) of amphiphilic colloidal particles. *Langmuir* **2008**, *24*, 2438–2445. [CrossRef] [PubMed]

48. Daware, S.V.; Basavaraj, M.G. Emulsions stabilized by silica rods via arrested demixing. *Langmuir* **2015**, *31*, 6649–6654. [CrossRef] [PubMed]

49. Schoth, A.; Landfester, K.; Munoz-Espi, R. Surfactant-free polyurethane nanocapsules via inverse Pickering miniemulsion. *Langmuir* **2015**, *31*, 3784–3788. [CrossRef] [PubMed]

50. Lou, F.; Ye, L.; Kong, M.; Yang, Q.; Li, G.; Huang, Y. Pickering emulsions stabilized by shape-controlled silica microrods. *RSC Adv.* **2016**, *6*, 24195–24202. [CrossRef]

51. De Folter, J.W.J.; Hutter, E.M.; Castillo, S.I.R.; Klop, K.E.; Philipse, A.P.; Kegel, W.K. Particle shape anisotropy in Pickering emulsions: Cubes and peanuts. *Langmuir* **2014**, *30*, 955–964. [CrossRef] [PubMed]

52. Ahn, W.J.; Jung, H.S.; Choi, H.J. Pickering emulsion polymerized smart magnetic poly(methyl methacrylate)/Fe_2O_3 composite particles and their stimulus-response. *RSC Adv.* **2015**, *5*, 23094–23100. [CrossRef]

53. Zhang, J.; Li, L.; Wang, J.; Xu, J.; Sun, D. Phase inversion of emulsions containing a lipophilic surfactant induced by clay concentration. *Langmuir* **2013**, *29*, 3889–3894. [CrossRef] [PubMed]

54. Yi, W.; Wu, H.; Wang, H.; Du, Q. Interconnectivity of macroporous hydrogels prepared via graphene oxide-stabilized Pickering high internal phase emulsions. *Langmuir* **2016**, *32*, 982–990. [CrossRef] [PubMed]

55. Creighton, M.A.; Zhu, W.; van Krieken, F.; Petteruti, R.A.; Gao, H.; Hurt, R.H. Three-dimensional graphene-based microbarriers for controlling release and reactivity in colloidal liquid phases. *ACS Nano* **2016**, *10*, 2268–2276. [CrossRef] [PubMed]

56. Creighton, M.A.; Ohata, Y.; Miyawaki, J.; Bose, A.; Hurt, R.H. Two-dimensional materials as emulsion stabilizers: Interfacial thermodynamics and molecular barrier properties. *Langmuir* **2014**, *30*, 3687–3696. [CrossRef] [PubMed]

57. Avendano, C.; Brun, N.; Fontaine, O.; In, M.; Mehdi, A.; Stocco, A.; Vioux, A. Multiwalled carbon nanotube/cellulose composite: From aqueous dispersions to Pickering emulsions. *Langmuir* **2016**, *32*, 3907–3916. [CrossRef] [PubMed]

58. Tasset, S.; Cathala, B.; Bizot, H.; Capron, I. Versatile cellular foams derived from CNC-stabilized Pickering emulsions. *RSC Adv.* **2014**, *4*, 893–898. [CrossRef]

59. Godfrin, M.P.; Tiwari, A.; Bose, A.; Tripathi, A. Phase and steady shear behavior of dilute carbon black suspensions and carbon black stabilized emulsions. *Langmuir* **2014**, *30*, 15400–15407. [CrossRef] [PubMed]

60. Hu, Z.; Marway, H.S.; Kasem, H.; Pelton, R.; Cranston, E.D. Dried and redispersible cellulose nanocrystal Pickering emulsions. *ACS Macro Lett.* **2016**, *5*, 185–189. [CrossRef]

61. Cherhal, F.; Cousin, F.; Capron, I. Structural description of the interface of Pickering emulsions stabilized by cellulose nanocrystals. *Biomacromolecules* **2016**, *17*, 496–502. [CrossRef] [PubMed]

62. Nishizawa, N.; Kawamura, A.; Kohri, M.; Nakamura, Y.; Fujii, S. Polydopamine particle as a particulate emulsifier. *Polymers* **2016**, *8*, 62. [CrossRef]

63. Qi, F.; Wu, J.; Sun, G.Q.; Nan, F.F.; Ngai, T.; Ma, G.H. Systematic studies of Pickering emulsions stabilized by uniform-sized PLGA particles: Preparation and stabilization mechanism. *J. Mater. Chem. B* **2014**, *2*, 7605–7611. [CrossRef]

64. Yang, Y.; Wei, Z.; Wang, C.; Tong, Z. Lignin-based Pickering HIPEs for macroporous foams and their enhanced adsorption of copper(II) ions. *Chem. Commun.* **2013**, *49*, 7144–7146. [CrossRef] [PubMed]

65. Zhang, Y.; Chen, Z.; Bian, W.; Feng, L.; Wu, Z.; Wang, P.; Zeng, X.; Wu, T. Stabilizing oil-in-water emulsions with regenerated chitin nanofibers. *Food Chem.* **2015**, *183*, 115–121. [CrossRef] [PubMed]

66. Tzoumaki, M.V.; Moschakis, T.; Kiosseoglou, V.; Biliaderis, C.G. Oil-in-water emulsions stabilized by chitin nanocrystal particles. *Food Hydrocoll.* **2011**, *25*, 1521–1529. [CrossRef]

67. Binks, B.P.; Isa, L.; Tyowua, A.T. Direct measurement of contact angles of silica particles in relation to double inversion of Pickering emulsions. *Langmuir* **2013**, *29*, 4923–4927. [CrossRef] [PubMed]

68. Worthen, A.J.; Foster, L.M.; Dong, J.; Bollinger, J.A.; Peterman, A.H.; Pastora, L.E.; Bryant, S.L.; Truskett, T.M.; Bielawski, C.W.; Johnston, K.P. Synergistic formation and stabilization of oil-in-water emulsions by a weakly interacting mixture of zwitterionic surfactant and silica nanoparticles. *Langmuir* **2014**, *30*, 984–994. [CrossRef] [PubMed]

69. Li, W.; Yu, L.; Liu, G.; Tan, J.; Liu, S.; Sun, D. Oil-in-water emulsions stabilized by Laponite particles modified with short-chain aliphatic amines. *Colloids Surf. A* **2012**, *400*, 44–51. [CrossRef]

70. Li, W.; Zhao, C.; Tan, J.; Jiang, J.; Xu, J.; Sun, D. Roles of methyl orange in preparation of emulsions stabilized by layered double hydroxide particles. *Colloids Surf. A* **2013**, *421*, 173–180. [CrossRef]

71. Santini, E.; Guzman, E.; Ferrari, M.; Liggieri, L. Emulsions stabilized by the interaction of silica nanoparticles and palmitic acid at the water-hexane interface. *Colloids Surf. A* **2014**, *460*, 333–341. [CrossRef]

72. Sadeghpour, A.; Pirolt, F.; Glatter, O. Submicrometer-sized Pickering emulsions stabilized by silica nanoparticles with adsorbed oleic acid. *Langmuir* **2013**, *29*, 6004–6012. [CrossRef] [PubMed]

73. Vilchez, A.; Rodriguez-Abreu, C.; Menner, A.; Bismarck, A.; Esquena, J. Antagonistic effects between magnetite nanoparticles and a hydrophobic surfactant in highly concentrated Pickering emulsions. *Langmuir* **2014**, *30*, 5064–5074. [CrossRef] [PubMed]

74. Sturzenegger, P.N.; Gonzenbach, U.T.; Koltzenburg, S.; Gauckler, L.J. Controlling the formation of particle-stabilized water-in-oil emulsions. *Soft Matter* **2012**, *8*, 7471–7479. [CrossRef]

75. Jiang, J.; Zhu, Y.; Cui, Z.; Binks, B.P. Switchable Pickering emulsions stabilized by silica nanoparticles hydrophobized in situ with a switchable surfactant. *Angew. Chem. Int. Ed.* **2013**, *52*, 12373–12376. [CrossRef] [PubMed]

76. Zhu, Y.; Jiang, J.; Liu, K.; Cui, Z.; Binks, B.P. Switchable Pickering emulsions stabilized by silica nanoparticles hydrophobized in situ with a conventional cationic surfactant. *Langmuir* **2015**, *31*, 3301–3307. [CrossRef] [PubMed]

77. Cui, Z.; Yang, L.; Cui, Y.; Binks, B.P. Effects of surfactant structure on the phase inversion of emulsions stabilized by mixtures of silica nanoparticles and cationic surfactant. *Langmuir* **2010**, *26*, 4717–4724. [CrossRef] [PubMed]

78. Cui, Z.; Cui, C.; Zhu, Y.; Binks, B.P. Multiple phase inversion of emulsions stabilized by in situ surface activation of $CaCO_3$ nanoparticles via adsorption of fatty acids. *Langmuir* **2012**, *28*, 314–320. [CrossRef] [PubMed]

79. Schrade, A.; Landfester, K.; Ziener, U. Pickering-type stabilized nanoparticles by heterophase polymerization. *Chem. Soc. Rev.* **2013**, *42*, 6823–6839. [CrossRef] [PubMed]

80. Williams, M.; Armes, S.P.; Verstraete, P.; Smets, J. Double emulsions and colloidosomes-in-colloidosomes using silica-based Pickering emulsifiers. *Langmuir* **2014**, *30*, 2703–2711. [CrossRef] [PubMed]

81. Hu, Z.; Patten, T.; Pelton, R.; Cranston, E.D. Synergistic stabilization of emulsions and emulsion gels with water-soluble polymers and cellulose nanocrystals. *ACS Sustain. Chem. Eng.* **2015**, *3*, 1023–1031. [CrossRef]

82. Wang, J.; Liu, G.; Wang, L.; Li, C.; Xu, J.; Sun, D. Synergistic stabilization of emulsions by poly(oxypropylene)diamine and Laponite particles. *Colloids Surf. A* **2010**, *353*, 117–124. [CrossRef]

83. Binks, B.P.; Murakami, R.; Armes, S.P.; Fujii, S. Temperature-induced inversion of nanoparticle-stabilized emulsions. *Angew. Chem. Int. Ed.* **2005**, *44*, 4795–4798. [CrossRef] [PubMed]

84. Yoon, K.Y.; Li, Z.; Neilson, B.M.; Lee, W.; Huh, C.; Bryant, S.L.; Bielawski, C.W.; Johnston, K.P. Effect of adsorbed amphiphilic copolymers on the interfacial activity of superparamagnetic nanoclusters and the emulsification of oil in water. *Macromolecules* **2012**, *45*, 5157–5166. [CrossRef]

85. Sun, G.; Li, Z.; Ngai, T. Inversion of particle-stabilized emulsions to form high-internal-phase emulsions. *Angew. Chem. Int. Ed.* **2010**, *49*, 2163–2166. [CrossRef] [PubMed]

86. Yang, H.; Zhou, T.; Zhang, W. A strategy for separating and recycling solid catalysts based on the pH-triggered Pickering-emulsion inversion. *Angew. Chem. Int. Ed.* **2013**, *52*, 7455–7459. [CrossRef] [PubMed]

87. Huang, J.; Yang, H. A pH-switched Pickering emulsion catalytic system: High reaction efficiency and facile catalyst recycling. *Chem. Commun.* **2015**, *51*, 7333–7336. [CrossRef] [PubMed]

88. Zhao, C.; Tan, J.; Li, W.; Tong, K.; Xu, J.; Sun, D. Ca^{2+} ion responsive Pickering emulsions stabilized by PSSMA nanoaggregates. *Langmuir* **2013**, *29*, 14421–14428. [CrossRef] [PubMed]

89. Lin, S.; Theato, P. CO_2-responsive polymers. *Macromol. Rapid Commun.* **2013**, *34*, 1118–1133. [CrossRef] [PubMed]

90. Liang, C.; Liu, Q.; Xu, Z. Surfactant-free switchable emulsions using CO_2-responsive particles. *ACS Appl. Mater. Interfaces* **2014**, *6*, 6898–6904. [CrossRef] [PubMed]

91. Fameau, A.L.; Lam, S.; Velev, O.D. Multi-stimuli responsive foams combining particles and self-assembling fatty acids. *Chem. Sci.* **2013**, *4*, 3874–3881. [CrossRef]

92. Chen, Z.; Zhou, L.; Bing, W.; Zhang, Z.; Li, Z.; Ren, J.; Qu, X. Light controlled reversible inversion of nanophosphor-stabilized Pickering emulsions for biphasic enantioselective biocatalysis. *J. Am. Chem. Soc.* **2014**, *136*, 7498–7504. [CrossRef] [PubMed]

93. Zhang, Q.; Bai, R.-X.; Guo, T.; Meng, T. Switchable Pickering emulsions stabilized by awakened TiO_2 nanoparticle emulsifiers using UV/dark actuation. *ACS Appl. Mater. Interfaces* **2015**, *7*, 18240–18246. [CrossRef] [PubMed]

94. Qian, Y.; Zhang, Q.; Qiu, X.; Zhu, S. CO_2-responsive diethylaminoethyl-modified lignin nanoparticles and their application as surfactants for CO_2/N_2-switchable Pickering emulsions. *Green Chem.* **2014**, *16*, 4963–4968. [CrossRef]

95. Tang, J.; Lee, M.F.X.; Zhang, W.; Zhao, B.; Berry, R.M.; Tam, K.C. Dual responsive Pickering emulsion stabilized by poly 2-(dimethylamino)ethyl methacrylate grafted cellulose nanocrystals. *Biomacromolecules* **2014**, *15*, 3052–3060. [CrossRef] [PubMed]

96. Tang, J.; Berry, R.M.; Tam, K.C. Stimuli-responsive cellulose nanocrystals for surfactant-free oil harvesting. *Biomacromolecules* **2016**, *17*, 1748–1756. [CrossRef] [PubMed]

97. Thompson, K.L.; Mable, C.J.; Lane, J.A.; Derry, M.J.; Fielding, L.A.; Armes, S.P. Preparation of Pickering double emulsions using block copolymer worms. *Langmuir* **2015**, *31*, 4137–4144. [CrossRef] [PubMed]

98. Yi, C.; Sun, J.; Zhao, D.; Hu, Q.; Liu, X.; Jiang, M. Influence of photo-cross-linking on emulsifying performance of the self-assemblies of poly(7-(4-vinylbenzyloxyl)-4-methylcoumarin-co-acrylic acid). *Langmuir* **2014**, *30*, 6669–6677. [CrossRef] [PubMed]

99. Mable, C.J.; Warren, N.J.; Thompson, K.L.; Mykhaylyk, O.O.; Armes, S.P. Framboidal ABC triblock copolymer vesicles: A new class of efficient Pickering emulsifier. *Chem. Sci.* **2015**, *6*, 6179–6188. [CrossRef]

100. Wei, W.; Wang, T.; Yi, C.; Liu, J.; Liu, X. Self-assembled micelles based on branched poly(styrene-alt-maleic anhydride) as particulate emulsifiers. *RSC Adv.* **2015**, *5*, 1564–1570. [CrossRef]

101. Yi, C.; Liu, N.; Zheng, J.; Jiang, J.; Liu, X. Dual-responsive poly(styrene-alt-maleic acid)-graft-poly(n-isopropyl acrylamide) micelles as switchable emulsifiers. *J. Colloid Interface Sci.* **2012**, *380*, 90–98. [CrossRef] [PubMed]

102. Chen, M.; Geng, Z.; Sun, M.; Liu, X.; Chen, S. Pickering emulsion stabilized by self-assembled micelles of amphiphilic random copolymer p(St-co-DM). *J. Dispers. Sci. Technol.* **2014**, *35*, 757–764. [CrossRef]

103. Fujii, S.; Cai, Y.L.; Weaver, J.V.M.; Armes, S.P. Syntheses of shell cross-linked micelles using acidic abc triblock copolymers and their application as pH-responsive particulate emulsifiers. *J. Am. Chem. Soc.* **2005**, *127*, 7304–7305. [CrossRef] [PubMed]

104. Ma, C.; Bi, X.; Ngai, T.; Zhang, G. Polyurethane-based nanoparticles as stabilizers for oil-in-water or water-in-oil Pickering emulsions. *J. Mater. Chem. A* **2013**, *1*, 5353–5360. [CrossRef]

105. Cunningham, V.J.; Armes, S.P.; Musa, O.M. Synthesis, characterisation and Pickering emulsifier performance of poly(stearyl methacrylate)-poly(n-2-(methacryloyloxy)ethyl pyrrolidone) diblock copolymer nano-objects via RAFT dispersion polymerisation in n-dodecane. *Polym. Chem.* **2016**, *7*, 1882–1891. [CrossRef]

106. Richtering, W. Responsive emulsions stabilized by stimuli-sensitive microgels: Emulsions with special non-pickering properties. *Langmuir* **2012**, *28*, 17218–17229. [CrossRef] [PubMed]

107. Destribats, M.; Eyharts, M.; Lapeyre, V.; Sellier, E.; Varga, I.; Ravaine, V.; Schmitt, V. Impact of PNIPAM microgel size on its ability to stabilize Pickering emulsions. *Langmuir* **2014**, *30*, 1768–1777. [CrossRef] [PubMed]

108. Morse, A.J.; Madsen, J.; Growney, D.J.; Armes, S.P.; Mills, P.; Swart, R. Microgel colloidosomes based on pH-responsive poly(tert-butylaminoethyl methacrylate) latexes. *Langmuir* **2014**, *30*, 12509–12519. [CrossRef] [PubMed]

109. Wang, W.; Milani, A.H.; Carney, L.; Yan, J.; Cui, Z.; Thaiboonrod, S.; Saunders, B.R. Doubly crosslinked microgel-colloidosomes: A versatile method for pH-responsive capsule assembly using microgels as macro-crosslinkers. *Chem. Commun.* **2015**, *51*, 3854–3857. [CrossRef] [PubMed]

110. Li, Z.; Ming, T.; Wang, J.; Ngai, T. High internal phase emulsions stabilized solely by microgel particles. *Angew. Chem. Int. Ed.* **2009**, *48*, 8490–8493. [CrossRef] [PubMed]

111. Li, Z.; Wei, X.; Ngai, T. Controlled production of polymer microspheres from microgel-stabilized high internal phase emulsions. *Chem. Commun.* **2011**, *47*, 331–333. [CrossRef] [PubMed]

112. Li, Z.; Ngai, T. Stimuli-responsive gel emulsions stabilized by microgel particles. *Colloid Polym. Sci.* **2011**, *289*, 489–496. [CrossRef]

113. Fujii, S.; Read, E.S.; Binks, B.P.; Armes, S.P. Stimulus-responsive emulsifiers based on nanocomposite microgel particles. *Adv. Mater.* **2005**, *17*, 1014–1018. [CrossRef]

114. Binks, B.P.; Murakami, R.; Armes, S.P.; Fujii, S. Effects of pH and salt concentration on oil-in-water emulsions stabilized solely by nanocomposite microgel particles. *Langmuir* **2006**, *22*, 2050–2057. [CrossRef] [PubMed]

115. Ngai, T.; Behrens, S.H.; Auweter, H. Novel emulsions stabilized by pH and temperature sensitive microgels. *Chem. Commun.* **2005**, 331–333. [CrossRef] [PubMed]

116. Brugger, B.; Richtering, W. Emulsions stabilized by stimuli-sensitive poly(N-isopropylacrylamide)-co-methacrylic acid polymers: Microgels versus low molecular weight polymers. *Langmuir* **2008**, *24*, 7769–7777. [CrossRef] [PubMed]

117. Brugger, B.; Rosen, B.A.; Richtering, W. Microgels as stimuli-responsive stabilizers for emulsions. *Langmuir* **2008**, *24*, 12202–12208. [CrossRef] [PubMed]

118. de Folter, J.W.J.; van Ruijven, M.W.M.; Velikov, K.P. Oil-in-water Pickering emulsions stabilized by colloidal particles from the water-insoluble protein zein. *Soft Matter* **2012**, *8*, 6807–6815. [CrossRef]

119. Fujii, S.; Aichi, A.; Muraoka, M.; Kishimoto, N.; Iwahori, K.; Nakamura, Y.; Yamashita, I. Ferritin as a bionano-particulate emulsifier. *J. Colloid Interface Sci.* **2009**, *338*, 222–228. [CrossRef] [PubMed]

120. Li, Z.; Xiao, M.; Wang, J.; Ngai, T. Pure protein scaffolds from Pickering high internal phase emulsion template. *Macromol. Rapid Commun.* **2013**, *34*, 169–174. [CrossRef] [PubMed]

121. Xiao, J.; Lu, X.; Huang, Q. Double emulsion derived from kafirin nanoparticles stabilized Pickering emulsion: Fabrication, microstructure, stability and in vitro digestion profile. *Food Hydrocoll.* **2017**, *62*, 230–238. [CrossRef]

122. Xiao, J.; Li, C.; Huang, Q. Kafirin nanoparticle-stabilized Pickering emulsions as oral delivery vehicles: Physicochemical stability and in vitro digestion profile. *J. Agric. Food Chem.* **2015**, *63*, 10263–10270. [CrossRef] [PubMed]

123. Liu, F.; Tang, C.-H. Soy protein nanoparticle aggregates as Pickering stabilizers for oil-in-water emulsions. *J. Agric. Food Chem.* **2013**, *61*, 8888–8898. [CrossRef] [PubMed]

124. Shimoni, G.; Shani Levi, C.; Levi Tal, S.; Lesmes, U. Emulsions stabilization by lactoferrin nano-particles under in vitro digestion conditions. *Food Hydrocoll.* **2013**, *33*, 264–272. [CrossRef]

125. Zou, Y.; Guo, J.; Yin, S.-W.; Wang, J.-M.; Yang, X.-Q. Pickering emulsion gels prepared by hydrogen-bonded zein/tannic acid complex colloidal particles. *J. Agric. Food Chem.* **2015**, *63*, 7405–7414. [CrossRef] [PubMed]

126. Gao, Z.M.; Yang, X.Q.; Wu, N.N.; Wang, L.J.; Wang, J.M.; Guo, J.; Yin, S.W. Protein-based Pickering emulsion and oil gel prepared by complexes of zein colloidal particles and stearate. *J. Agric. Food Chem.* **2014**, *62*, 2672–2678. [CrossRef] [PubMed]

127. Juttulapa, M.; Piriyaprasarth, S.; Sriamornsak, P. Effect of pH on stability of oil-in-water emulsions stabilized by pectin-zein complexes. In *Multi-Functional Materials and Structures IV*; Sombatsompop, N., Bhattacharyya, D., Cheung, K.H.Y., Eds.; Trans Tech Publications Ltd.: Stafa-Zurich, Switzerland, 2013; Volume 747, pp. 127–130.

128. Destribats, M.; Rouvet, M.; Gehin-Delval, C.; Schmitt, C.; Binks, B.P. Emulsions stabilised by whey protein microgel particles: Towards food-grade Pickering emulsions. *Soft Matter* **2014**, *10*, 6941–6954. [CrossRef] [PubMed]

129. Meshulam, D.; Lesmes, U. Responsiveness of emulsions stabilized by lactoferrin nano-particles to simulated intestinal conditions. *Food Funct.* **2014**, *5*, 65–73. [CrossRef] [PubMed]

130. Gonzalez-Jordan, A.; Nicolai, T.; Benyahia, L. Influence of the protein particle morphology and partitioning on the behavior of particle-stabilized water-in-water emulsions. *Langmuir* **2016**, *32*, 7189–7197. [CrossRef] [PubMed]

131. Binks, B.P.; Fletcher, P.D.I. Particles adsorbed at the oil-water interface: A theoretical comparison between spheres of uniform wettability and "Janus" particles. *Langmuir* **2001**, *17*, 4708–4710. [CrossRef]

132. Tu, F.; Lee, D. One-step encapsulation and triggered release based on Janus particle-stabilized multiple emulsions. *Chem. Commun.* **2014**, *50*, 15549–15552. [CrossRef] [PubMed]

133. Mejia, A.F.; Diaz, A.; Pullela, S.; Chang, Y.-W.; Simonetty, M.; Carpenter, C.; Batteas, J.D.; Mannan, M.S.; Clearfield, A.; Cheng, Z. Pickering emulsions stabilized by amphiphilic nano-sheets. *Soft Matter* **2012**, *8*, 10245–10253. [CrossRef]

134. Cao, Z.; Wang, G.; Chen, Y.; Lang, F.; Yang, Z. Light-triggered responsive Janus composite nanosheets. *Macromolecules* **2015**, *48*, 7256–7261. [CrossRef]

135. Kim, J.W.; Cho, J.; Cho, J.; Park, B.J.; Kim, Y.-J.; Choi, K.-H.; Kim, J.W. Synthesis of monodisperse bi-compartmentalized amphiphilic Janus microparticles for tailored assembly at the oil-water interface. *Angew. Chem. Int. Ed.* **2016**, *55*, 4509–4513. [CrossRef] [PubMed]

136. Xu, F.; Fang, Z.; Yang, D.; Gao, Y.; Li, H.; Chen, D. Water in oil emulsion stabilized by tadpole-like single chain polymer nanoparticles and its application in biphase reaction. *ACS Appl. Mater. Interfaces* **2014**, *6*, 6717–6723. [CrossRef] [PubMed]

137. Yi, F.; Xu, F.; Gao, Y.; Li, H.; Chen, D. Macrocellular polymer foams from water in oil high internal phase emulsion stabilized solely by polymer Janus nanoparticles: Preparation and their application as support for Pd catalyst. *RSC Adv.* **2015**, *5*, 40227–40235. [CrossRef]

138. Chen, Y.; Liang, F.; Yang, H.; Zhang, C.; Wang, Q.; Qu, X.; Li, J.; Cai, Y.; Qiu, D.; Yang, Z. Janus nanosheets of polymer-inorganic layered composites. *Macromolecules* **2012**, *45*, 1460–1467. [CrossRef]

139. Fujii, S.; Yokoyama, Y.; Miyanari, Y.; Shiono, T.; Ito, M.; Yusa, S.-I.; Nakamura, Y. Micrometer-sized gold–silica Janus particles as particulate emulsifiers. *Langmuir* **2013**, *29*, 5457–5465. [CrossRef] [PubMed]

140. Zhao, Z.; Liang, F.; Zhang, G.; Ji, X.; Wang, Q.; Qu, X.; Song, X.; Yang, Z. Dually responsive Janus composite nanosheets. *Macromolecules* **2015**, *48*, 3598–3603. [CrossRef]

141. Ruhland, T.M.; Groschel, A.H.; Ballard, N.; Skelhon, T.S.; Walther, A.; Muller, A.H.E.; Bon, S.A.F. Influence of Janus particle shape on their interfacial behavior at liquid-liquid interfaces. *Langmuir* **2013**, *29*, 1388–1394. [CrossRef] [PubMed]

142. Kim, J.-W.; Lee, D.; Shum, H.C.; Weitz, D.A. Colloid surfactants for emulsion stabilization. *Adv. Mater.* **2008**, *20*, 3239–3243. [CrossRef]

143. Yamagami, T.; Kitayama, Y.; Okubo, M. Preparation of stimuli-responsive "mushroom-like" Janus polymer particles as particulate surfactant by site-selective surface-initiated AGET ATRP in aqueous dispersed systems. *Langmuir* **2014**, *30*, 7823–7832. [CrossRef] [PubMed]

144. Tu, F.; Lee, D. Shape-changing and amphiphilicity-reversing Janus particles with pH-responsive surfactant properties. *J. Am. Chem. Soc.* **2014**, *136*, 9999–10006. [CrossRef] [PubMed]

materials

MDPI

Article

Transition Behaviors of Configurations of Colloidal Particles at a Curved Oil-Water Interface

Mina Lee, Ming Xia and Bum Jun Park *

Department of Chemical Engineering, Kyung Hee University, Yongin, Gyeonggi-do 17104, Korea;
mina.lee@khu.ac.kr (M.L.); mxia19861123@gmail.com (M.X.)
* Correspondence: bjpark@khu.ac.kr; Tel.: +82-31-201-2429

Academic Editor: To Ngai
Received: 28 January 2016; Accepted: 24 February 2016; Published: 26 February 2016

Abstract: We studied the transition behaviors of colloidal arrangements confined at a centro-symmetrically curved oil-water interface. We found that assemblies composed of several colloidal particles at the curved interface exhibit at least two unique patterns that can be attributed to two factors: heterogeneity of single-colloid self-potential and assembly kinetics. The presence of the two assembly structures indicates that an essential energy barrier between the two structures exists and that one of the structures is kinetically stable. This energy barrier can be overcome via external stimuli (e.g., convection and an optical force), leading to dynamic transitions of the assembly patterns.

Keywords: colloids; fluid-fluid interfaces; assemblies; interactions; heterogeneity; self-potentials

1. Introduction

Colloidal particles strongly and irreversibly attach to fluid-fluid interfaces [1–3]. This adsorption leads to a reduction in the interfacial tension between the immiscible fluid phases, thereby causing thermodynamic or kinetic stabilization of the interface [1,3,4]. Interactions between interface-trapped particles are abnormally strong and long-ranged, which differs significantly from DLVO (Derjaguin-Landau-Verwey-Overbeek) interactions where identical particles are dispersed in a single fluid phase, such as water [5–18]. In addition to such interparticle interactions, assembly/microstructures and rheological properties of colloidal particles confined at fluid-fluid interfaces have been extensively investigated over the last three decades; this is because such particle-laden interface systems can be used for various applications, such as solid surfactants (e.g., Pickering emulsions), material transfer/delivery, and microemulsion catalytic reactors [19–29].

Control over the assembly of colloidal particles with versatile functionalities at fluid-fluid interfaces can be used to establish micro- and nano-scale building blocks [2]. Consequently, the study of small-scale measurements (e.g., pair interactions, assemblies, and micromechanics) can provide a direct link to measurements (e.g., micro- and macro-structures and rheology) on a larger scale for materials that are composed of the same types of constituents, thereby potentially offering fundamental ideas and design rules for the construction of hierarchical structures and materials with specific properties. For example, two particle interactions with a few microns in size at a planar oil-water interface were measured via optical laser tweezers, and the magnitude of the obtained interaction forces was found to depend on the particle pairs (*i.e.*, the pair interaction heterogeneity) [29]. Based on Monte Carlo (MC) simulations and experimental observations, it was found that the measured interaction heterogeneity of the particles directly affects the conformation of the equilibrium microstructure of two-dimensional colloidal suspensions consisting of identical particles. Similarly, it was reported that the interactions between polystyrene particles with a few hundred micrometers in diameter at a centro-symmetrically curved oil-water interface are repulsive, which can affect the formation of diverse assembly patterns composed of several particles [30]. The diversity in the assembly structures can be attributed to heterogeneity in interparticle interactions. However, it should be noted that the magnitude of the interaction potentials for each interacting particle, rather than the pair interaction magnitude, is heterogeneous, and thus, the heterogeneity of the potentials of individual particles likely affects their assembly behaviors.

The heterogeneous properties in colloidal interaction systems are crucial because the assumption of particle interaction homogeneity can result in a deviation from predicted large-scale properties for systems composed of identical particles. Such variations could occur in the structural and rheological properties of a suspension. To understand the heterogeneity of colloidal systems on a single particle level, a new concept of self-potentials possessed by individual particles trapped at an oil-water interface has been developed [31]. To characterize the self-potentials of individual colloidal particles, energy minimization is performed numerically when the interface-trapped particles form uniquely arranged structures. Notably, it was demonstrated that the self-potentials represent the dipole strength of individual particles at the interface, which can account for electrostatic dipolar interactions with abnormally strong and long-ranged properties [5–9].

In this work, we focus on studying transition behaviors of assembly configurations composed of several particles at a curved oil-water interface, based on the self-potential model. We experimentally observed that configurations of polystyrene particles at the curved interface that are repulsive to each other can be varied between two different structures by introducing external stimuli (convection and optical forces). This implies that one of the structures is kinetically stable and the other is thermodynamically stable. We believe that these structural transitions of the particle assemblies are due to the effects of particle interaction heterogeneity and assembly kinetics. We use MC simulations to systematically understand the role of the two factors, employing the heterogeneity of the self-potentials. This paper is organized as follows. First, we present experimental observations of the assembly pattern formation and the structural transition induced by convective flow or laser irradiation. Then, we present data from MC simulations to quantitatively investigate the effect of the interaction heterogeneity and assembly kinetics on the assembly behaviors. Finally, we describe the detailed experimental procedure for the preparation of particles and the formation of assembly patterns at the curved oil-water interface. This is followed by a description of the simulation method.

2. Results and Discussion

2.1. Experimental Observations of Assembly Configurations

A convex oil lens was formed by gently placing a small amount of *n*-decane on the surface of ultrapure water in a Petri dish (Figure 1). Polystyrene particles with 200 µm in diameter (2*R*) were individually inserted to a curved decane-water interface in which the curvature of the interface can be determined by solving the nonlinear Yong-Laplace equation [23,32]. The particles trapped at the interface tend to collect near the bottom of the oil lens due to gravity and also experience electrostatic repulsions [5–9], forming unique assembly structures. We refer the readers to the Section 3 for the detailed experimental methods.

It was observed that the assembly composed of eight to ten particles at the curved oil-water interface shows two unique patterns depending on the number of particles. The assembly of eight particles adopts either the configuration with one inner particle surrounded by seven outer particles (1@7) or the structure with two inner particles surrounded by six outer particles (2@6). Nine and ten particle assemblies show either two inside structures (2@7 and 2@8) or three inside structures (3@6 and 3@7), respectively. We believe that these assembly behaviors can be attributed to heterogeneity of self-potentials. The role of the self-potential heterogeneity that is incorporated with MC simulations is investigated in the next section.

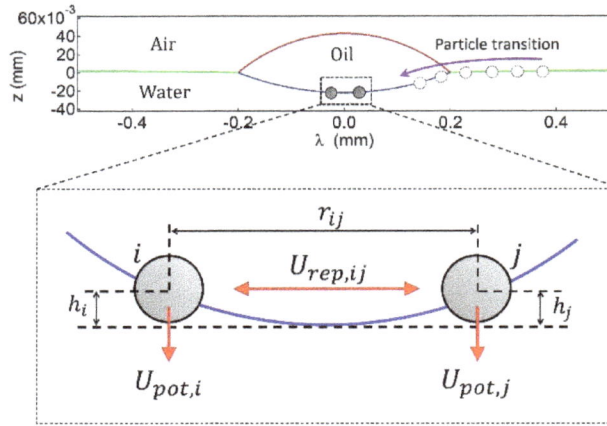

Figure 1. Schematics for particle interactions at the curved oil-water interface.

The assembly behaviors also likely depend on a kinetic factor. A particle is added to the 2@6 assembly to experimentally observe the effect of assembly kinetics on the particle configurations. As shown in Figure 2a, as the added particle approaches along the direction of the line joining the two inner particles, it joins the outer particles forming the 2@7 structure. In contrast, when the inserted particle approaches the 2@6 structure at the angle of the line joining the two inner particles (Figure 2b), the nine particles consequently adopt the 3@6 structure. These experimental results indicate that the kinetic factor plays an important role in the formation of assembly structures.

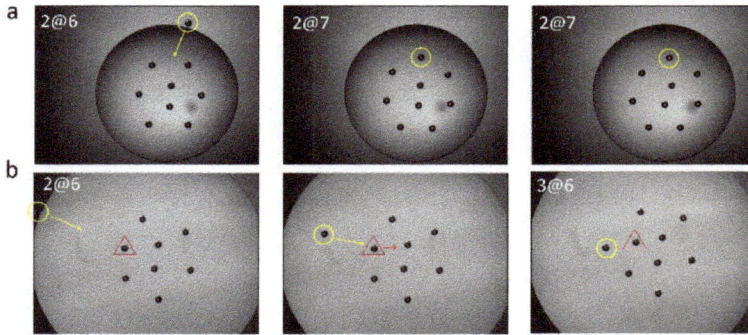

Figure 2. Experimental snapshots of the structure transitions when a ninth particle is added to the eight particle assembly: (**a**) from 2@6 to 2@7; and (**b**) from 2@6 to 3@6; the scale bar is 1mm.

Interestingly, the assembly pattern can be transferred to a different structure via an external stimulus, such as convection due to oil drying. The sample cell containing the particles attached to the curved oil-water interface is uncovered under ambient conditions, allowing the oil to dry, enabling convection to occur. As shown in Figure 3, the assembly pattern consisting of nine particles at the interface initially shows two inner particles surrounded by seven outer particles (2@7). Upon oil drying, one of the outer particles enters inward to form an inner triangle enclosed by six outer particles (3@6) after a period of time. Since this structural change seldom occurs when the sample cell is covered and convection is suppressed, we believe that convection caused by oil drying leads to such a structural transition.

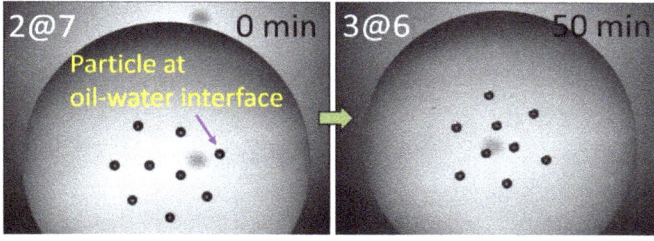

Figure 3. Structural transition form 2@7 to 3@6 assembly patterns at the curved oil-water upon oil drying; the scale bar is 1mm.

We also observed that the assembly configurations consisting of eight and eleven particles can also be altered by stimulating the structure via an optical force. Note that eleven particles also form two distinct patterns, either 3@8 or 4@7 structures. A laser beam (power ≈ 1 mW, wavelength = 650 nm) is introduced from a side port of the microscope, which reaches the focal plane through the aperture of a microscope objective. The rays of the laser beam are transmitted and reflected at the particle-fluid interface, as well as the fluid-fluid interfaces, generating an optical force that can randomly disturb the assembly structure [33]. Note that the sample cell is covered to minimize the effect of convection due to evaporating oil. As shown in Figure 4a, the assembly composed of eight particles initially exhibits a 2@6 structure. After approximately 15 min of irradiation with the laser beam, one of the inner particles (indicated by a red circle) moves outward to form the 1@7 structure. This configuration is maintained for a while, even after the removal of the laser beam. Further irradiation for another 15 min leads to the recovery of the original structure. Similarly, when eleven particles initially form the structure in which four particles are surrounded by seven outer particles (4@7), laser irradiation for ~15 min leads to a transition into the 3@8 structure, as shown in Figure 4b. The reverse transition (back to the original configuration), occurs upon exposure to the laser beam for another 15 min. In short, these observed experimental results likely demonstrate the presence of an energy barrier between the two assembly structures that can be overcome by an external stimulus, and one of the assembly structures should be kinetically stable. Next, we use MC simulations to systematically understand these assembly behaviors.

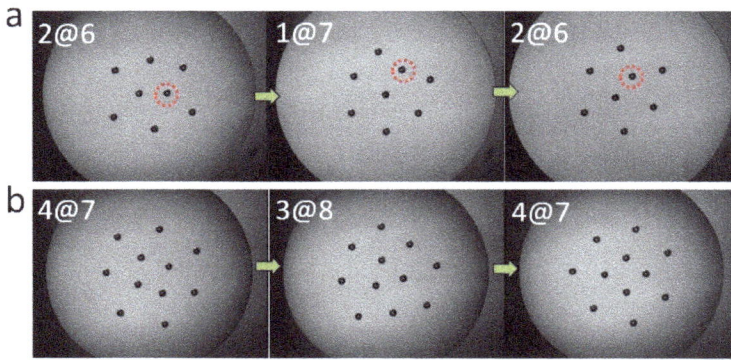

Figure 4. Transition of the assembly pattern activated by an optical force: (**a**) between 2@6 and 1@7; and (**b**) between 4@7 and 3@8; the scale bar is 1mm.

2.2. Assembly Behaviors via MC Simulations

There are two factors that likely determine the assembly structures: interaction heterogeneity and assembly kinetics. To understand the effects of interaction heterogeneity, we run MC simulations by introducing the heterogeneity of self-potentials that were previously measured [31].

When self-potentials for the nine particles are randomly allocated for each run of the simulation, as described earlier, the particles form the 3@6 structure 34% of the time (68 out of 200 runs) and the 2@7 structure 66% of the time. To perform a more quantitative analysis, we select an arbitrary set of self-potentials for the nine particles, as shown in Figure 5d. The initial positions of the nine particles at the curved oil-water interface in the simulations are also randomly generated. Multiple runs of simulations show that the nine particles form the 3@6 structure 62% of the time (124 out of 200 runs). Among the simulation results that form the 3@6 structure, the particles with the three lowest self-potentials (particles 6, 8, and 9) almost always collect inward (~90% of the time). It is likely that the particles with lower self-potentials tend to possess the larger number of nearest neighbors and, as a result, collect inside of the structure. Among the remaining simulation results that assemble into the 2@7 structure (76 out of 200), particles 6 and 8, which possess the two lowest self-potentials, form the two inner particles 32% of the time. Additionally, at least one particle among the six or eight is always located inside the 2@7 pattern. Notably, when a homogeneous interparticle interaction is introduced to the simulations, the nine particles only form the 3@6 structure (100 out of 100). These simulation results demonstrate the importance of interaction heterogeneity in assembly behavior.

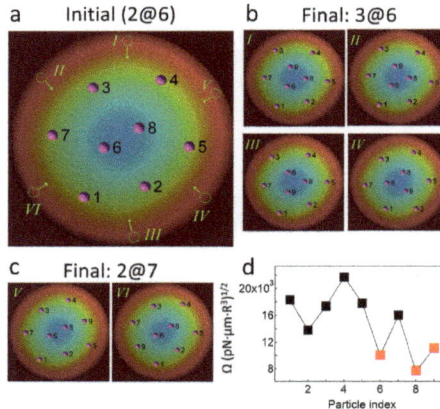

Figure 5. Path-dependent structural transition behavior. (**a**) The initial configuration of the 2@6 structure for the MC simulations. Particle 9 approaches the 2@6 structure form six locations, as indicated by the dotted circles (I-VI); (**b,c**) The configurations resulting from each run of the MC simulations; and (**d**) The self-potentials (Ω_i) for the nine particles used in the MC simulations.

The assembly kinetics also significantly affects the formation of assembly patterns. We investigate the effect of the path of a particle that is added to the 2@6 assembly structure using MC simulations. We arbitrarily select a 2@6 configuration as the initial state for the simulations (Figure 5a) in which the self-potential values for the eight particles in Figure 5d are randomly selected among the regenerated self-potential values based on the gamma distribution, as described in the Section 3. Then, particle 9, which possesses the third-lowest self-potential in Figure 5d, is inserted into the equilibrium 2@6 structure from six locations (I-VI), as indicated by the dotted circles in Figure 5a. Note that this simulation geometry is analogous to the experimental conditions in Figure 2. As shown in Figure 5c, the 2@7 structure is obtained when the inserted particle approaches along the path that is aligned with the line joining the two inner particles, particles 6 and 8 (cases V and VI in Figure 5c). In the other cases, where particle 9 approaches the 2@6 structure at the angle of the line joining the two inner particles, the nine particles consequently form the 3@6 structure (cases I-IV in Figure 5b). Note that when forming the 3@6 structure, the indexes of the inner particles are always 6, 8, and 9, which possess the three lowest self-potentials.

Further simulations (500 runs) are performed by adding particle 9 to the 2@6 structure, in which the initial position of particle 9 is arbitrarily located on the circular boundary, as indicated by the dashed circle in Figure 6a. The probability of the two resulting assembly patterns forming is indicated by various colors on the initial positions of the circular boundary in Figure 6b. Consistently, a higher probability (red) for the 3@6 formation is obtained when particle 9 approaches from the approximate directions of either 1-2 or 7-8 o'clock; for the other cases, the 2@7 assembly is preferred. Note that the probability (colors on the circular boundary in Figure 6b) of forming two unique structures (2@7 and 3@6) is not symmetric with respect to the line joining the two inner particles, particles 6 and 8, due to the heterogeneity of the self-potentials of the nine particles (Figure 5d). Interestingly, when one particle is added to the 2@6 configuration, the corresponding probabilities for the 3@6 and 2@7 structures to form are 71% and 29%, respectively. These differ from the probabilities in the case where the random initial configurations and the same self-potentials are used in the simulations (62% for 3@6 and 38% for 2@7). These simulation results, which depend on the initial configurations as well as the path of an inserted particle, demonstrate the effect of kinetics on determining the assembly patterns.

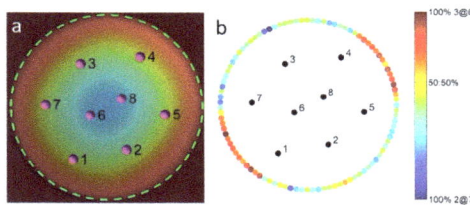

Figure 6. The probability of the formation of either the 2@7 or 3@6 structures upon the addition of particle 9 to the 2@6 structure. (**a**) Initial 2@7 configuration. The dashed circle represents the initial position of the inserted particle; (**b**) The probability of the two resulting structures is indicated by various colors on the initial position.

The total interaction potential for the formation of the 3@6 and the 2@7 structures continuously decreases until particle 9 finds a position at which a local energy minimum likely exists, as shown in Figure 7. The magnitude of the total interaction potential for the 3@6 formation is ~2×10^7 k_BT lower than that of the 2@7 formation, suggesting that the 2@7 formation is a kinetically stable configuration. Similar results are obtained when either particle 6 or 8 is inserted into the 2@6 assembly (instead of particle 9) in the simulations.

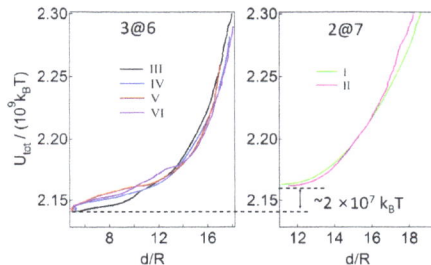

Figure 7. The total interaction energy (U_{tot}) as a function of the radial distance, d, of particle 9 from the center of the oil lens while rearrangements occur.

To understand the effects of the self-potential magnitude on the assembly configurations, we determine the critical value of the self-potentials to be the point for which the assembly structure transitions to a different structure. The three inner particles of the 3@6 structure typically possess the three lowest self-potentials, as described earlier, implying that a critical value of the self-potential of the inner particles should exist. To find this critical magnitude, the self-potential of one of the three inner particles (Figure 5d) is amplified in the MC simulations until the particle migrates outward, leading to the formation of the 2@7 structure (Figure 8a,b). It is found that particle 8 transfers to the outside when its self-potential is scaled by a factor of up to $\chi = 1.7$, such that $\Omega_8' = 1.70 \times \Omega_8 \approx 1.30 \times 10^4$ $(pN\mu mR^3)^{1/2}$. At the moment of amplifying the self-potential of particle 8, the transient energy field with respect to particle 8 (Figure 8e) shows that the location of particle 8 in the 3@6 structure deviates slightly from the local energy minimum (inset in Figure 8e). This, consequently, displaces the particle in an outward direction to decrease the total energy until the particle reaches the position of the local energy minimum in the 2@7 structure (Figure 8f). Similarly, the reverse transition occurs when the amplified factor of particle 8 is removed ($\chi = 1.0$), such that $\Omega_8' = \Omega_8$. As shown in the transient energy field (Figure 8g), particle 8 in the 2@7 structure moves inward due to deviations in the particle position caused by the transient local energy minimum. Eventually, particle 8 joins the two inner particles 6 and 9, and assembles into the 3@6 pattern (Figure 8h). The particles' trajectories and the total interaction energies during the structural transitions are shown in Figure 8c,d, respectively. The solid symbols in Figure 8c indicate the initial positions of the particles before each transition occurs. Similarly, the critical self-potentials for particles 6 and 9 are found to be $\Omega_6' = 1.56 \times \Omega_6 \approx 1.56 \times 10^4$ $(pN\mu mR^3)^{1/2}$ and $\Omega_9' = 1.17 \times \Omega_9 \approx 1.30 \times 10^4$ $(pN\mu mR^3)^{1/2}$, respectively. The simulation results indicate that the transition tends to occur when the critical self-potentials are comparable to the value of the fourth lowest self-potential (the value of particle 2). In addition, upon replacing the value of Ω_8 with the critical value of $\Omega_8' = 1.30 \times 10^4$ $(pN\mu mR^3)^{1/2}$, we run MC simulations with random initial configurations. The multiple runs of simulations show that the probability of forming the 3@6 structure dramatically decreases (7% of the time), compared to the case of the simulations when the original values of self-potentials in Figure 5d are used (62%). The increase in the self-potential value for particle 8 leads to the formation of the preferred 2@7 structure. In short, the simulation results demonstrate that the assembly structures depend on the relative strength of the self-potentials and the relative positions of the surrounding inner and outer particles.

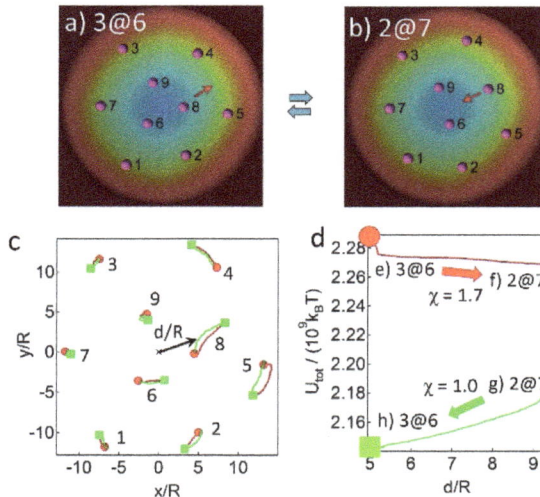

Figure 8. *Cont.*

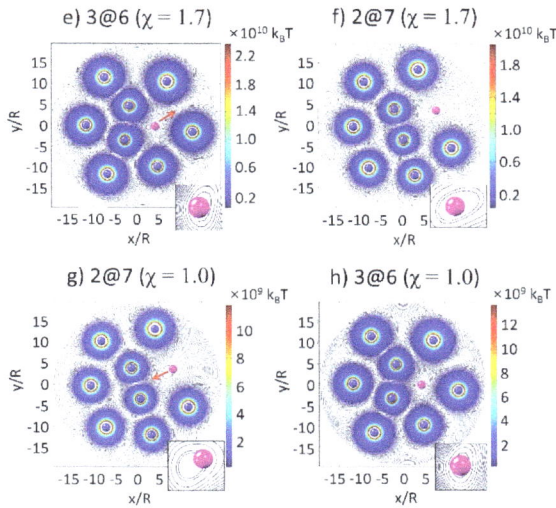

Figure 8. Magnitude of the critical self-potentials at which the structural transition occurs between the 3@6 and 2@7 patterns in MC simulations. The self-potential values in Figure 5d are used in the simulations. (**a,b**) The structural transition when the critical self-potential of particle 8, $\Omega_8' = 1.70 \times \Omega_8 \approx 1.30 \times 10^4$ $(pN\mu m R^3)^{1/2}$, is used in the simulation; (**c,d**) The corresponding particles' trajectories; (**c**) and the total interaction energies; (**d**) when the transition between the 3@6 and 2@7 structures occurs; (**e–h**) The transition energy field (contour lines) at the moment of the structural transition with respect to particle 8. The pink dot indicates the position of particle 8 and the blue dots are the other eight particles. The contour lines near each blue particle is not shown because the potential is extremely high.

Finally, it is likely that the pattern formation depends on the history of sequential addition of particles that belongs to the effect of assembly kinetics. One by one addition of particles to the curved interface, for example, results in assembly patterns of triangle, quadrangle, 1@4, 1@5, 1@6, 2@6, 3@6, and 3@7, as shown in the experimental and simulation results in Figure 9. The similar pattern formation, however, can also be observed when the same number of particles are simultaneously added to the interface. Therefore, we believe that it is not trivial to solely extract the role of the sequential addition because the two effects (*i.e.*, interaction heterogeneity and assembly kinetics) simultaneously affect the assembly behaviors of particles at the curved interface.

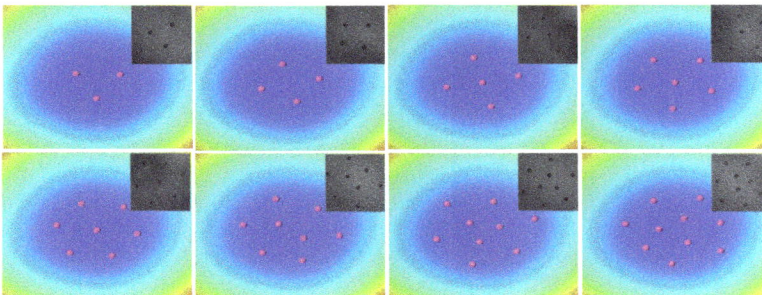

Figure 9. Examples of assembly patterns obtained by simulations and experiments when particles are added to the curved oil-water interface in a sequential manner.

3. Materials and Methods

3.1. Preparation of Particles

Polystyrene (PS) particles that are 200 μm in diameter ($2R$) were prepared using a microfluidic device with co-flow geometry [23]. 30 wt% polystyrene (M.W. = 190 K, Sigma-Aldrich, Yongin, Korea) in methylene chloride (Sigma-Aldrich, Yongin, Korea) forms droplets out of a tapered inner glass capillary tube, and the outer continuous phase is water containing 2 wt.% polyvinyl alcohol (PVA, Sigma-Aldrich, Yongin, Korea), which prevents coalescence of the droplets. The generated droplets were collected in a pure aqueous solution and dried at an ambient temperature for several days. This particle dispersion was washed with pure water several times to remove excess PVA. To measure the surface potential of the PS particles, 30 wt.% PS dissolved in methylene chloride was dispersed in 2 wt.% PVA using an ultrasonic cleaner (Fisher Scientific, Seoul, Korea). The methylene chloride in the dispersed phase was removed at an ambient temperature for several days and the resulting particle dispersion was washed with pure water several times. The ζ-potential was then measured at approximately −20 mV (Beckman Coulter Delsa Nano-C, Indianapolis, IN, USA) due to the presence of PVA moieties on the particle surface that can consequently lead to repulsive interparticle interactions at an oil-water interface [23].

3.2. Formation of Convex Oil Lens

To generate a centro-symmetrically-curved oil-water interface, a small amount of *n*-decane is placed on the surface of water (resistivity > 18.2 MΩ·cm) in a Petri dish (Falcon, Yongin, Korea). A convex oil lens is formed and its shape (*i.e.*, air-water, air-oil, and oil-water interfaces in Figure 1) can be determined as a function of the radial distance, λ, via the Mathematica Player file provided by Burton *et al.*, [32] in which the parameter values are the surface tensions of the air-oil ($\gamma_{ao} \approx 23.8$ mN/m) and air-water ($\gamma_{aw} \approx 72$ mN/m) interfaces, the spreading coefficient ($S_{decane} = -3.72 \pm 0.24$), and the density of decane ($\rho = 730$ kg/m^3).

3.3. Particle Adsorption to the Curved Oil-Water Interface

Particles were individually inserted into the air-water interface outside the oil lens using a micropipette. The transition of a particle from the air-water interface to the centro-symmetrically curved oil-water interface occurs spontaneously, minimizing the attachment energy of the particle to the fluid interfaces (air-water, air-oil, and oil-water), as well as the potential energy due to gravity (Figure 1) [23]. To place several particles at the curved oil-water interface, a number of particles are initially spread at the planar air-water interface and then a small amount of oil is added onto the water surface, allowing the particles to transition to the curved oil-water interface. After introducing the particles, the sample cell is covered to minimize convection, if necessary. To observe a convection-induced structural change in the particles, the sample cell is uncovered, leading to evaporation of the oil. Note that since the bond number (the ratio of the gravitational force to the interfacial tension force) is sufficiently small (Bo $\approx 10^{-3}$), the local interface deformation around the particle is negligible. Microscopic snapshots are captured by a charge-coupled device (CCD) camera (Hitachi KP-M1AN, Daejeon, Korea) in order to image the assembly configurations. The obtained images are analyzed with ImageJ software [34].

3.4. Monte Carlo Simulation

We use Monte Carlo (MC) simulations to quantitatively analyze the assembly behaviors at the curved oil-water interface. N particles at the interface repel each other due to electrostatic repulsion, and the corresponding pair interaction between particles i and j is given by:

$$\frac{U_{rep,ij}}{k_B T} = \frac{\Omega_i \Omega_j}{r_{ij}^3} = \frac{a_{ij}}{r_{ij}^3} \tag{1}$$

where r_{ij} is the particle separation; k_B is the Boltzmann's constant; and T is the temperature [6,18,30]. It has been reported that the electrostatic repulsive interactions between particles at fluid-fluid interfaces (*i.e.*, oil-water and air-water interfaces) scale as r^{-3} [5–9]. Such abnormally strong and long-ranged repulsive interactions stem from the dipole-dipole interactions due to asymmetric charge distribution across the interface [5–7] and/or the presence of the surface residual charges in the apolar phase (oil or air) [8,9]. The power law exponent of the repulsive interaction at the fluid-fluid interface ($U_{rep} \sim r^{-3}$) has also been demonstrated by the direct measurements using the optical laser tweezers [8,18,23,29]. The particles i and j possess their own potentials, the self-potentials, Ω_i and Ω_j, respectively [31]. Notably, it was previously found that the measured values of self-potentials are heterogeneous, and that the random combination of the self-potentials corresponds to the pair interaction magnitude ($a_{ij} = \Omega_i \Omega_j$) and follows a gamma distribution:

$$f(a; k, \theta) = a^{k-1} \frac{e^{-a/\theta}}{\theta^k \Gamma(k)} \tag{2}$$

where the shape and scale parameters are $k = 4.36$; and $\theta = 5.48 \times 10^7$ pNμmR^3, respectively; and $\Gamma(k)$ is the gamma function [31]. To introduce the heterogeneity of self-potential to MC simulations, the values of self-potentials for a sufficiently large number of particles (500) are regenerated in the range of the measured self-potentials. Then, a set of self-potentials for N particles is randomly selected among the regenerated self-potential values for further simulation studies.

N particles tend to collect near the bottom due to gravity (Figure 1). The corresponding potential energy, upon the incorporation of buoyancy, can be expressed as follows:

$$\frac{U_{pot}}{k_B T} = g\left(\rho_p V_p - \rho_w V_{pw} - \rho_o V_{po}\right) \sum_{k=1}^{N} h_k \tag{3}$$

where V_p is the volume of a particle; ρ is the density; and g is the acceleration due to gravity [30]. The vertical distance between the k-th particle and the bottom of the oil lens, h_k, is determined by interpolating the interface height at the radial distance (λ) of the particle from the center of the oil lens (Figure 1). The particle volumes immersed in the water and oil phases are given as:

$$V_{pw} = \frac{\pi R^3}{3}\left(2 + 3\cos\theta_c - \cos^3\theta_c\right) \tag{4}$$

in water and

$$V_{po} = \frac{\pi R^3}{3}\left(2 - 3\cos\theta_c + \cos^3\theta_c\right) \tag{5}$$

in oil. The gel-trapping method is used to measure the three-phase contact angle of a particle ($\theta_c = 105°$) at the oil-water interface [30,35].

Assuming pairwise additivity [17], the net interaction energy of particle i is the sum of all pair interactions with its surrounding particles j; that is:

$$U_{net,i} = U_{pot,i} + \sum_{j,j \neq i}^{N} U_{rep,ij} \tag{6}$$

Notably, the pairwise additivity of the interaction potentials was previously justified when the interactions between the particles are sufficiently strong and long-ranged, such as the electrostatic repulsions between particles at fluid-fluid interfaces [17]. The total energy of an assembly composed of N particles at the curved oil-water interface is then given by:

$$U_{tot} = \sum_{i}^{N} U_{net,i} \tag{7}$$

The total energy is used to determine whether to accept or reject a new configuration after particle i is randomly moved. Initial positions of the number of particles in MC simulations are randomly generated at the curved oil-water interface. The size of the oil lens does not significantly affect the assembly behaviors [31]; therefore, its radius is arbitrarily chosen to be R_{lens} = 2 mm.

4. Conclusions

The assembly behavior of colloidal particles at the curved oil-water interface has been investigated. Based on the experimental observations and MC simulations, assembly behaviors of several particles at the curved oil-water interface can be summarized as follows. (1) The values of self-potential and its heterogeneity directly affect the assembly behaviors (*i.e.*, the diverse assembly pattern formation); (2) the formation of assembly structure also depends on the assembly kinetics; (3) the presence of the two structures demonstrates that one of the structures is kinetically stable and that there is an essential corresponding energy barrier between the two structures. The two structures can be transformed from one to the other upon the introduction of external stimuli, such as convection or an optical force, which likely imparts the activation energy required to overcome the energy barrier. Notably, this assembly behavior can be typically applied for cases when the particles assemble a maximum two different structures ($N \leqslant 20$). Based on preliminary results after performing more MC simulations, more than two different assembly structures can occur for the larger number of particle systems, suggesting the possibility of increasing the number of kinetically stable assembly patterns; and (4) a very interesting feature observed in the assembly patterns is that the particles with low values of self-potentials likely form inner particles and the particles with relatively high values of self-potentials stay outside, such that the resulting structure composed of several particles is energetically favorable. Based on this assembly behavior, it may be possible to sort particles in the order of their self-potential values by subjecting sufficiently strong external energy to the sample such as mechanical vibration and, therefore, the particles that are radially distributed at the curved interface can adopt a thermodynamically stable structure.

In addition, interactions (electrostatic repulsions) between colloidal particles at a fluid-fluid interface do not significantly depend on the interfacial curvature when the curvature radius is sufficiently large in comparison to the particle size; however, the assembly structures and behaviors at a curved interface are notably different from those at a planar interface. For a planar interface, when homogeneous repulsive interaction potentials are employed in the simulations, the resulting assembly structures should show a hexagonal honeycomb pattern with six neighboring particles. In contrast, the introduction of heterogeneous potentials to the simulations results in an assembly which includes colloidal defects with particles possessing five or seven neighboring particles [29]. The assembly structures at a curved interface necessarily possess colloidal defects, regardless of the introduction of either heterogeneous or homogeneous potentials to the simulations. This is analogous to a football, which cannot be organized only with hexagons. Therefore, this work has the potential to provide a fundamental basis from which to understand the intimate relationship between small-scale measurements such as interparticle interactions and assembly behaviors, and the large-scale properties of materials composed of many identical particles such as micro- and macro-structures and interfacial rheology.

Acknowledgments: This work was supported by grants from Kyung Hee University (KHU-20141582), the Basic Science Research Program of the National Research Foundation (NRF) of Korea funded by the Ministry of Science, ICT & Future Planning (MSIP) (NRF-2014R1A1A1005727), and the Engineering Research Center of Excellence Program of Korea funded by the MSIP/NRF of Korea (NRF-2014R1A5A1009799).

Author Contributions: Mina Lee and Ming Xia performed the experiments and contributed in the discussion. Bum Jun Park planned the experiments, performed the MC simulations, and wrote the manuscript.

Conflicts of Interest: The authors declare no conflict of interest.

References

1. Binks, B.P.; Horozov, T.S. *Colloidal Particles at Liquid Interfaces*; Cambridge University Press: New York, NY, USA, 2006.

2. Park, B.J.; Lee, D. Particles at fluid-fluid interfaces: From single-particle behavior to hierarchical assembly of materials. *MRS Bulletin* **2014**, *39*, 1089–1098. [CrossRef]
3. Aveyard, R. Can Janus particles give thermodynamically stable Pickering emulsions? *Soft Matter* **2012**, *8*, 5233–5240. [CrossRef]
4. Binks, B.P. Particles as surfactants-similarities and differences. *Curr. Opin. Colloid Interface Sci.* **2002**, *7*, 21–41. [CrossRef]
5. Pieranski, P. Two-dimensional interfacial colloidal crystals. *Phys. Rev. Lett.* **1980**, *45*, 569–572. [CrossRef]
6. Hurd, A.J. The electrostatic interaction between interfacial colloidal particles. *J. Phys. A Math. Gen.* **1985**, *18*. [CrossRef]
7. Oettel, M.; Dietrich, S. Colloidal interactions at fluid interfaces. *Langmuir* **2008**, *24*, 1425–1441. [CrossRef] [PubMed]
8. Aveyard, R.; Binks, B.P.; Clint, J.H.; Fletcher, P.D.; Horozov, T.S.; Neumann, B.; Paunov, V.N.; Annesley, J.; Botchway, S.W.; Nees, D.; *et al.* Measurement of long-range repulsive forces between charged particles at an oil-water interface. *Phys. Rev. Lett.* **2002**, *88*. [CrossRef] [PubMed]
9. Aveyard, R.; Clint, J.H.; Nees, D.; Paunov, V.N. Compression and structure of monolayers of charged latex particles at air/water and octane/water interfaces. *Langmuir* **2000**, *16*, 1969–1979. [CrossRef]
10. Botto, L.; Lewandowski, E.P.; Cavallaro, M.; Stebe, K.J. Capillary interactions between anisotropic particles. *Soft Matter* **2012**, *8*, 9957–9971. [CrossRef]
11. Cavallaro, M., Jr.; Botto, L.; Lewandowski, E.P.; Wang, M.; Stebe, K.J. Curvature-driven capillary migration and assembly of rod-like particles. *Proc. Natl. Acad. Sci. USA* **2011**, *108*, 20923–20928. [CrossRef] [PubMed]
12. Danov, K.D.; Kralchevsky, P.A.; Naydenov, B.N.; Brenn, G. Interactions between particles with an undulated contact line at a fluid interface: Capillary multipoles of arbitrary order. *J. Colloid Interface Sci.* **2005**, *287*, 121–134. [CrossRef] [PubMed]
13. Kralchevsky, P.A.; Nagayama, K. Capillary forces between colloidal particles. *Langmuir* **1994**, *10*, 23–36. [CrossRef]
14. Lewandowski, E.P.; Cavallaro, M.; Botto, L.; Bernate, J.C.; Garbin, V.; Stebe, K.J. Orientation and self-assembly of cylindrical particles by anisotropic capillary interactions. *Langmuir* **2010**, *26*, 15142–15154. [CrossRef] [PubMed]
15. Loudet, J.C.; Alsayed, A.M.; Zhang, J.; Yodh, A.G. Capillary interactions between anisotropic colloidal particles. *Phys. Rev. Lett.* **2005**, *94*. [CrossRef] [PubMed]
16. Madivala, B.; Fransaer, J.; Vermant, J. Self-assembly and rheology of ellipsoidal particles at interfaces. *Langmuir* **2009**, *25*, 2718–2728. [CrossRef] [PubMed]
17. Park, B.J.; Lee, B.; Yu, T. Pairwise interactions of colloids in two-dimensional geometric confinement. *Soft Matter* **2014**, *10*, 9675–9680. [CrossRef] [PubMed]
18. Park, B.J.; Pantina, J.P.; Furst, E.M.; Oettel, M.; Reynaert, S.; Vermant, J. Direct measurements of the effects of salt and surfactant on interaction forces between colloidal particles at water-oil interfaces. *Langmuir* **2008**, *24*, 1686–1694. [CrossRef] [PubMed]
19. Ballard, N.; Bon, S.A.F. Hybrid biological spores wrapped in a mesh composed of interpenetrating polymer nanoparticles as "patchy" Pickering stabilizers. *Polym. Chem.* **2011**, *2*, 823–827. [CrossRef]
20. Dinsmore, A.D.; Hsu, M.F.; Nikolaides, M.G.; Marquez, M.; Bausch, A.R.; Weitz, D.A. Colloidosomes: Selectively permeable capsules composed of colloidal particles. *Science* **2002**, *298*, 1006–1009. [CrossRef] [PubMed]
21. Kim, J.-W.; Larsen, R.J.; Weitz, D.A. Synthesis of nonspherical colloidal particles with anisotropic properties. *J. Am. Chem. Soc.* **2006**, *128*, 14374–14377. [CrossRef] [PubMed]
22. Kim, J.-W.; Lee, D.; Shum, H.C.; Weitz, D.A. Colloid surfactants for emulsion stabilization. *Adv. Mater.* **2008**, *20*, 3239–3243. [CrossRef]
23. Park, B.J.; Lee, D. Spontaneous particle transport through a triple-fluid phase boundary. *Langmuir* **2013**, *29*, 9662–9667. [CrossRef] [PubMed]
24. Pickering, S.U. Emulsions. *J. Chem. Soc. Trans.* **1907**, *91*, 2001–2021. [CrossRef]
25. Ramsden, W. Separation of solids in the surface-layers of solutions and 'suspensions' (observations on surface-membranes, bubbles, emulsions, and mechanical coagulation).—Preliminary account. *Proc. R. Soc. London* **1903**, *72*, 156–164. [CrossRef]

26. Reynaert, S.; Moldenaers, P.; Vermant, J. Control over colloidal aggregation in monolayers of latex particles at the oil-water interface. *Langmuir* **2006**, *22*, 4936–4945. [CrossRef] [PubMed]

27. Stancik, E.J.; Widenbrant, M.J.O.; Laschitsch, A.T.; Vermant, J.; Fuller, G.G. Structure and dynamics of particle monolayers at a liquid-liquid interface subjected to extensional flow. *Langmuir* **2002**, *18*, 4372–4375. [CrossRef]

28. Crossley, S.; Faria, J.; Shen, M.; Resasco, D.E. Solid nanoparticles that catalyze biofuel upgrade reactions at the water/oil interface. *Science* **2010**, *327*, 68–72. [CrossRef] [PubMed]

29. Park, B.J.; Vermant, J.; Furst, E.M. Heterogeneity of the electrostatic repulsion between colloids at the oil-water interface. *Soft Matter* **2010**, *6*, 5327–5333. [CrossRef]

30. Lee, M.; Lee, D.; Park, B.J. Effect of interaction heterogeneity on colloidal arrangements at a curved oil-water interface. *Soft Matter* **2015**, *11*, 318–323. [CrossRef] [PubMed]

31. Lee, M.; Park, B.J. Heterogeneity of single-colloid self-potentials at an oil-water interface. *Soft Matter* **2015**, *11*, 8812–8817. [CrossRef] [PubMed]

32. Burton, J.C.; Huisman, F.M.; Alison, P.; Rogerson, D.; Taborek, P. Experimental and numerical investigation of the equilibrium geometry of liquid lenses. *Langmuir* **2010**, *26*, 15316–15324. [CrossRef] [PubMed]

33. Park, B.J.; Furst, E.M. Optical trapping forces for colloids at the oil-water interface. *Langmuir* **2008**, *24*, 13383–13392. [CrossRef] [PubMed]

34. Schneider, C.A.; Rasband, W.S.; Eliceiri, K.W. NIH Image to ImageJ: 25 years of image analysis. *Nat. Meth.* **2012**, *9*, 671–675. [CrossRef]

35. Paunov, V.N. Novel method for determining the three-phase contact angle of colloid particles adsorbed at air-water and oil-water interfaces. *Langmuir* **2003**, *19*, 7970–7976. [CrossRef]

Communication

Assembly and Rearrangement of Particles Confined at a Surface of a Droplet, and Intruder Motion in Electro-Shaken Particle Films

Zbigniew Rozynek [1,*], Milena Kaczmarek-Klinowska [1] and Agnieszka Magdziarz [2]

[1] Faculty of Physics, Adam Mickiewicz University, Umultowska 85, Poznań 61-614, Poland; mkacz@amu.edu.pl

[2] Institute of Physical Chemistry, Polish Academy of Sciences, Kasprzaka 44/52, Warsaw 01-224, Poland; amagdziarz@ichf.edu.pl

* Correspondence: zbigniew.rozynek@gmail.com; Tel.: +48-503-775401

Academic Editor: To Ngai
Received: 8 June 2016; Accepted: 5 August 2016; Published: 10 August 2016

Abstract: Manipulation of particles at the surface of a droplet can lead to the formation of structures with heterogeneous surfaces, including patchy colloidal capsules or patchy particles. Here, we study the assembly and rearrangement of microparticles confined at the surface of oil droplets. These processes are driven by electric-field-induced hydrodynamic flows and by 'electro-shaking' the colloidal particles. We also investigate the motion of an intruder particle in the particle film and present the possibility of segregating the surface particles. The results are expected to be relevant for understanding the mechanism for particle segregation and, eventually, lead to the formation of new patchy structures.

Keywords: assembly; intruder motion; segregation of particles; electro-shaking; particle film; patchy structures; patchy colloidal capsules

---

1. Introduction

In this article, we study the assembly and rearrangement of colloidal particles confined at the surface of oil droplets, and we also communicate the possibility of utilizing electric fields for the segregation of particles at a surface of a droplet. A method for segregation or separation of the surface particles would likely be useful in fabricating patchy structures, particularly patchy colloidal capsules.

Patchy structures comprise at least two components with different functionalities; therefore, they possess interesting properties. For example, owing to specific interactions between patches, patchy structures can either self-assemble into complex structures or specifically adhere to a surface [1–3], guided–align [4–6], or self–propel [7–9]. Patchy capsules can additionally be used for storage, transportation, and release of cargo species [10]. There are dozens of methods for fabricating patchy particles [11–15], and some of those methods allow fast production of patchy particles in large quantities. In contrast, a large-scale fabrication of patchy colloidal capsules remains challenging; therefore, researchers seek for new methods of efficient production. Thus far, just a few routes to preparing patchy colloidal capsules have been demonstrated, including production by means of microfluidics [5,10,16] and mechanical pipetting [17–19]. In both of those, a patchy colloidal capsule is made via the coalescence of two or more droplets, each with different types of particles. It would be more effective if patchy colloidal capsules were made without the need of pairing the droplets and coalescing. Bulk emulsification is the approach that may offer higher material throughput at lower costs, as compared to the abovementioned methodologies. The challenge is to segregate the different particles contained in each droplet of an emulsion. Within this article, we made an attempt to manipulate and eventually segregate different particles located at the surface of one droplet, as a step needed for the formation of patchy colloidal structures.

Manipulation of particles bounded to a surface of a droplet (hereafter called surface particles) can be achieved by employing electric fields. The family of electrokinetic phenomena offers different types of mechanisms for transportation of the surface particles. The motion of particles can be induced either by electric forces that act directly on a particle (e.g., dielectrophoresis [20] or electrophoresis [21]) or by convective flows of liquids that are produced by electric fields [19]. These mechanisms can be used not only for transportation but also for segregation of the surface particles. For a droplet with two kinds of surface particles (a binary mixture), the dielectrophoresis (DEP) can be utilized to separate these particles: particles that may differ either by size or dielectric properties [21,22]. It was also shown that two types of particles located at the surface of a droplet can be separated the synergetic action of DEP and electrohydrodynamic (EHD) flows [23].

Here, we present experimental results that indicate the possibility for segregating particles by electric fields employing EHD flows and electro-shaking of the particle film. In the first part of this work, we further investigate the effect of the electro-shaking of particle film (the mechanism that was partially described in [24]) and provide new experimental results on the amplitude of the surface particle motion as a function of both the electric field strength and its frequency.

In the second part we present the study of motion of individual particle (called the intruder) travelling through the electro-shaken particle film. The intruder particles differ by size (is larger) from the particles comprising the particle film. The particle film has a ring-shape and is initially formed at the surface of the droplet by electric-field-assisted convective assembly [19]. The particle film is composed of densely packed (i.e., nearly hexagonal packing) particles. For allowing the intruder particle to move through the particle film, we electro-shake the particle film causing unjamming of particles forming the film. Thus, the intruder particles can successively (during each electro-shaking cycle) move through the particle film.

In the last part, we present preliminary experiments on particle segregation. For that, we use a dispersion of particles of two different sizes. We observe the segregation of particle, though it takes long time (many electro-shaking cycles) to achieve this.

2. Results

2.1. Formation of a Particle Film

In this research we study microparticles that are located at the surface of a droplet, where they are bound to the oil–oil interface by capillary force. As a silicone droplet is initially formed in castor oil by pipetting, most of the particles are located inside the droplet, and thus need to be brought to the surface of the droplet. This can be done by any suitable method: for example, by gentle mechanical stirring or by utilizing particle sedimentation due to the density mismatch between the particles and the silicone oil [20,24]. Once all the particles are adsorbed at the surface of the droplet, we assemble them at the electric equator of the droplet to form a monolayered ring-like structure (see Figure 1 and Movie S1). This is done by the action of the EHD flows induced by an E-field, as described in [19]. The surface particles form a densely ordered particle film. The particle film remains stable as long as the intensity of the E-field does not exceed the critical value; above the critical value, instabilities occur and the ring-like structure breaks apart, forming domains that spin [19], and at very high E-fields the particles are removed from the surface of the droplet into the castor oil (see Figure S1). When the E-field is turned off, the surface particles may slowly disintegrate from the film starting from those at the boundary of the film (the film either completely 'liquefies' or defragments), or it can stay stable; and the behavior of the particle film depends on the types of particles that compose the particle film. We observe that the attractive interactions between the polystyrene (PS) particles themselves are very small, though noticeable, i.e., some of the particles stick to one another, forming short chains (few to several particles long) aligned along the ring, which would be perpendicular to the direction of the

E-field; and, thus, the particle film defragments rather than completely liquefies. The particle film may remain stable if the attractive interactions between particles are strong, and for example, clay mineral particles adhere to one another rather strongly via tiny water bridges, and such a clay particle film is stable even if the *E*-field is turned off [24].

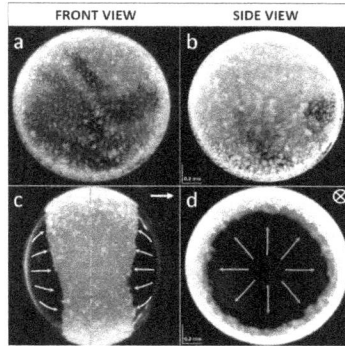

Figure 1. Droplet of silicone oil of diameter ~1.3 mm with PS10 particles. The particles are randomly distributed at the surface of the droplet before the *E*-field is applied (**a,b**). After application of the *E*-field of about 250 Vmm^{-1}, the particles are guided towards the 'electric equator' of the droplet by the EHD flows. The droplet is imaged perpendicular (**a,c**) and parallel (**b,d**) to the direction of *E*-field. See also corresponding Movie S1.

2.2. Electric-Field-Induced 'Shaking' of Particles

Shaking of the particle film can be induced by changing polarity of an electric field, and the motion of particles is predominantly in the horizontal direction (along the electric field direction). The mechanism of 'electro-shaking' is partially described in [24]. In short, by changing polarity of an *E*-field, a droplet may undergo a shape transition, from being oblate (flatted at the electric poles) to prolate (elongated along the electric poles) and then back to the oblate shape; this occurs due to a redistribution of charges at the surface of the droplet [24]. The particles move slightly upwards or downwards along the ring because the length of the equator changes as the droplet changes its shape. We note that because the PS particles' dielectric properties are similar to those of the oils, we do not observe particle alignment due to the particle-particle dipolar interactions (within the range of *E*-field strengths used here), as was the case when we used other particles, such as silver spheres [4] or clay minerals [25,26].

If a droplet is placed midway between the electrodes, the EHD flows at the surface of the droplet are roughly symmetrical in respect to the plane of symmetry that bisects the droplet and the particle film in a direction perpendicular to the *E*-field lines, as depicted by the blue dashed line in Figure 1c. Consequently, the induced motions of the electro-shaken particles are roughly symmetric towards/outwards the ring—that is, the particles in the middle of the film move the least, and the particles at either film boundaries move with the highest amplitudes. However, if the droplet is placed at a side of the cell and touches one of the electrodes, the flows of liquid are greatly suppressed on the side where the droplet touches the electrode, hence the EHD flows are not symmetrical anymore. Now, the particles that oscillate with the least amplitude are those at the particle film boundary on the side of the droplet that touches the electrode, whereas the particles on the other side of the film (along the width of the particle film) will have the highest amplitude of motion. In the experiments presented here, we work with a droplet touching one electrode, and there are two reasons for choosing this system: (i) it gives us more control of the droplet that ultimately stays in one location during the experiment; and (ii) we avoid any additional movement of surface particles that may occur because of the motion of the droplet.

The particle behavior during the electro-shaking resembles that of the granular media confined in a two-dimensional cell [27–30]. Generally, at very low electric fields, the particles within the film remain well ordered and do not relocate. As the intensity of the electric field increases, the particle film undergoes small defragmentation. Further increase of the intensity of the E-field causes the film to 'liquefy' (though some particles may stick to one another to form chains, which also undergo the motion), allowing the surface particles to relocate. At high E-fields, the amplitude of particle oscillation is very large. However, at the compression phase (i.e., at $\varphi \to n \cdot 180°$) the compressed particle film starts to deform (crumple), and at very high E-fields, the surface particles may eventually irreversibly detach from the surface of the droplet (see Figure S1).

An example of one cycle of the electro-shaking of particles at the surface of the droplet is presented in Figure 2. The AC E-field (~650 Vmm^{-1}, 0.5 Hz) is vertical, whereas the direction of gravity is horizontal. At $\varphi = 0°$ the particle film is densely packed, whereas at $\varphi = 90°$ it acquires the loosest state, and then at $\varphi = 180°$, the film is packed again.

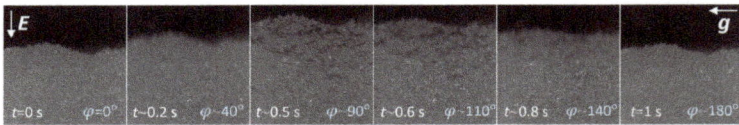

Figure 2. An example of one cycle of the electro-shaking. The AC E-field (~650 Vmm^{-1}, 0.5 Hz) is in a vertical direction, whereas the direction of gravity is horizontal. See also corresponding Movie S2.

Figure 3a shows the amplitude of particle oscillation A, (calculated as the difference of the positions of the particles at the boundary of the film, at $\varphi = 0$ and $\varphi = 90$, respectively) versus the intensity of the electric field, E, and its frequency, f. In order to find out how A scales with E and f, we normalized the data by $(f/f_R)^{-4}$, where $f_R = 1$ Hz is the arbitrarily chosen reference value. The overall data (which collapsed onto each other after normalization) follows the dashed line with a slope of 3.5 (see Figure 3b). Thus, for a droplet sized around 2 mm and with a particle film width of around 0.8 mm and the tested ranges of frequencies (0.5–1 Hz) and strengths of E-field (160–800 Vmm^{-1}), the amplitude of particle oscillation scales as $A \propto E^{3.5} \cdot f^{-4}$.

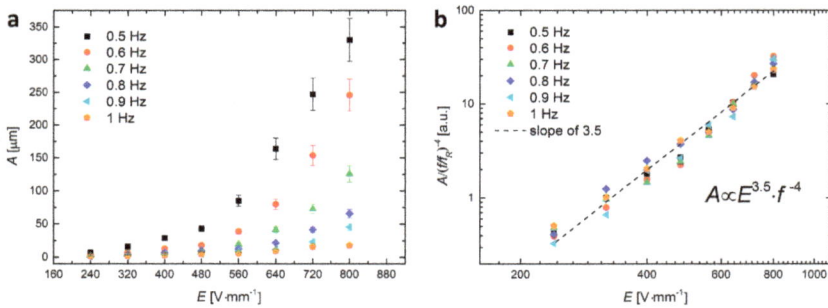

Figure 3. (a) The amplitude of particle oscillation versus the intensity of the electric field and its frequency; (b) A log–log plot of the data normalized by $(f/f_R)^{-4}$, where $f_R = 1$ Hz is the arbitrarily chosen reference value. The collapsed data roughly follows the dashed line with a slope of 3.5; thus, $A \propto E^{3.5} \cdot f^{-4}$ within the tested ranges of frequency and intensity of E-field.

Using frequencies higher than 1 Hz is impractical here, since the efficiency of electro-shaking is dropping. This is because at each event of changing polarity of the E-field, the droplet needs to charge up and deform. This takes at least 1.2 s for the system of liquids used here [17]. For frequencies lower than 0.5 Hz, and at high intensities of E-field, the particle film crumples and particles may detach from the surface of the droplet, as mentioned earlier.

2.3. An Intruder Travelling through the Particle Film

A single particle (hereafter called the intruder) of radius R was introduced into a sea of roughly monodispersed smaller particles of radius r (where $R/r \approx 14$) via electro-shaking. Initially, the intruder was located at the boundary of the particle film. In order to lodge the intruder particle at the film boundary, we electro-coalesced two silicone oil droplets, one with already prepared particle film and the other smaller droplet with an intruder particle situated on the surface of that droplet. The resulting coalesced droplet was subjected to the electric field (square-shaped) of around 600 Vmm^{-1} and with a frequency of 0.5 Hz.

Figure 4a shows snapshots from the experiment in which the PS140 intruder particle, initially located at the boundary of the film composed of PS10 particles, travels through the film along the direction of shaking, towards the electrode (top part of each image). Figure 4b shows the height of the intruder versus the number of shaking cycles. The data shows that the velocity of the intruder particle moving through the particle film is fastest at the beginning and slowest at the end of the measurement. When the data is plotted in the log–log scale (see the inset of Figure 4b) the measurement points roughly follow the red line with a slope of 0.5; hence, the distance travelled by the intruder scales as $N^{0.5}$.

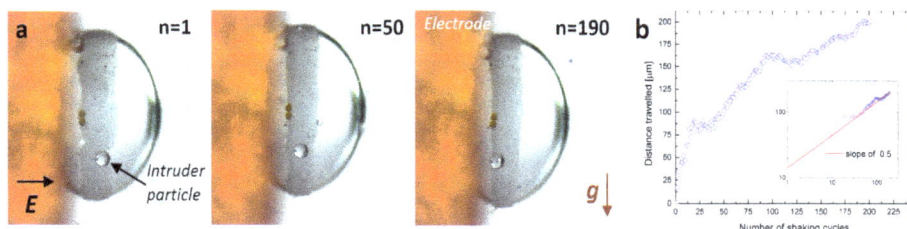

Figure 4. (**a**) Snapshots from the experiment with the PS140 intruder particle travelling through the PS10 film during the electro-shaking. Images taken after 1, 50, and 190 cycles of shaking, from left to right, respectively. Electric field of 600 Vmm^{-1} and 0.5 Hz was in a horizontal direction, whereas the gravitational force was in a vertical direction. See also corresponding Movie S3; (**b**) the distance travelled by the intruder versus the number of shaking cycles. The log–log plot is presented in the inset. The measurement points roughly follow the red line with a slope of 0.5; hence, we may conclude that the data scales as $N^{0.5}$.

Similar experiments were performed at the DC E-field. For the electric field strengths at which the particle film is stable, the intruder particle did not move through the particle film, even if the experiment lasted several minutes. Thus, the motion of the intruder particles was only possible when the particle film was temporarily unjammed due to the electro-shaking.

2.4. Size-Segregation of Surface Particles

Figure 5 presents snapshots from the experiment on size-segregating of the surface particles. We prepared a dispersion of PS10 and PS140 particles in silicone oil (with much larger amount of PS10 particles than PS140), and formed a droplet in castor oil. All the steps leading to formation of a ring-shaped particle film were the same as in all above-presented experiments. Before we began to electro-shake (E ~600 Vmm^{-1}, 0.5 Hz) the particle film, five PS140 particles had been distributed randomly within the film. Images shown in Figure 5 were taken after 1, 30, 90, and 200 cycles of electro-shaking, from left to right, respectively. After about 90 cycles all the PS140 particles were moved to the boundary of the film. Further electro-shaking did not change much in the system.

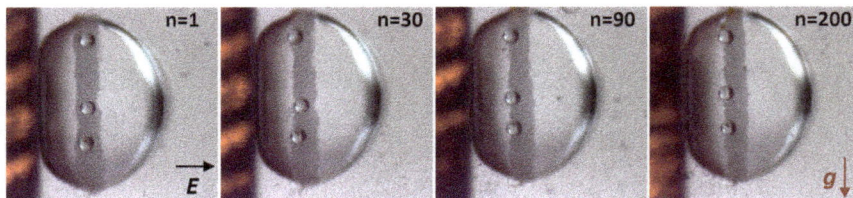

Figure 5. Snapshots from the experiment, in which we attempted to size-segregate the surface particles. Five PS140 particles were initially distributed randomly within the film composed of PS10 particles. Images taken after 1, 30, 90, and 200 cycles of electro-shaking, from left to right, respectively. After about 90 cycles all the PS140 particles were moved to the boundary of the film. Electric field of ~600 Vmm^{-1} and 0.5 Hz was in a horizontal direction, whereas the gravitational force was in a vertical direction. See also corresponding Movie S4.

3. Discussion

We presented experimental results that indicate the possibility for segregating particles by electric fields employing EHD flows and electro-shaking of the particle film. We studied motion of intruder particles travelling through the electro-shaken particle film composed of particles with smaller size than that of the intruder particles. For allowing the intruder particle to move through the particle film, we were electro-shaking the particle film to unjam particles forming the film and let the intruder particles to pass through the film. We note that the gravitational force was always in a direction perpendicular to the direction of the motion of the intruder particle. Thus, we conclude that the gravity plays no role in the mechanism of the motion of the intruder particle. Possibly the motion of the intruder particle is due to the electric force (e.g., dielectrophoretic or electrophoretic force) acting directly on particles, which is very small in comparison to the force stemming from the EHD flows. Further investigations are needed to provide with a firm understanding of the mechanism of the motion of the intruder particle. Nevertheless, within this communication, we highlight the possibilities of manipulating and eventually segregating different particles located at the surface of a droplet, which can be essential in the process of the formation of patchy colloidal structures. Thus, our results can open new pathways towards the formation of new architectures and structures, such as patchy colloidal capsules or patchy particles. The latter can be produced by solidification of the core liquid after the desired heterogeneous pattern is formed at the surface of the droplet.

4. Materials and Methods

The experimental set-up consisted of a sample cell placed on a mechanical x-y-z translational stage, with two digital microscopes for front and side viewing (perpendicular and parallel to the electric-field direction, respectively), a signal generator, a voltage amplifier, an oscilloscope for monitoring signal shape and amplitude, and a PC for collecting images. The sample cell was made of glass with two of the walls coated with a conductive indium tin oxide (ITO) layer, constituting electrodes. The high-voltage bipolar signal was provided to the cell via two crocodile clips attached to the ITO electrodes. The cell was filled with castor oil (83912, Sigma-Aldrich, St. Louis, MO, USA, density of 0.961 gcm^{-3} at 25 °C, electrical conductivity of around 50 pSm^{-1}, relative permittivity 4.7, and viscosity of around 700 cSt). A silicone oil droplet containing polystyrene (PS) particles was made in the castor oil using a mechanical pipette (see Figure S2a). In order to minimize the buoyancy force on the droplet with particles, two silicone oils with densities 0.960 and 0.965 gcm^{-3} at 25 °C (200/10 cSt and 200/100 cSt, Dow Corning, Auburn, AL, USA, electrical conductivity approximately 3–5 pSm^{-1}, relative permittivity 2.1) were mixed adequately to match the castor oil density. Two batches of the PS particles (PS10 and PS140), with diameters of around 10 μm and 140 μm and specific density of around 1 gcm^{-3}, were purchased from Microbeads AS, Skedsmokorset, Norway. The particles were surface modified to change their affinity towards castor oil, which resulted in different contact angles at the castor oil–silicone oil interface, hence influencing the stability of the particles at the surface of the droplet. The acrylate polymer (PFC 502AFA, FluoroPEL™, Cytonix, Beltsville, MD, USA) was used for the surface modification, and methoxy-nonafluorobutane, (7100 Engineered Fluid, 3M™ Novec™, St. Paul, MN, USA) was used as a solvent. The modification steps were as follows: (1) a solution of acrylate polymer and methoxy-nonafluorobutane solvent was prepared in concentration 1:300; (2) the PS particles (20 g) were dispersed in the solution (40 mL); (3) the solvent was removed by using a rotary evaporator, first at 10 min at 50 °C and afterwards at 10 min at 80 °C; (4) the modified PS particles were thoroughly washed with the solvent; and (5) the solvent residues were removed again using the rotary evaporator. The results of such modification are presented in Figure S2b. The high-voltage electric signal was obtained by amplifying a low-voltage signal (SDG1025 Siglent, Nashville, TN, USA) using a high-voltage bipolar amplifier (10HVA24-BP1 HVP, Planegg, Germany). The AC electric signal was always square-shaped and bipolar, and its value was provided as the RMS value (i.e., half of the peak-to-peak value).

Supplementary Materials: The following are available online at www.mdpi.com/1996-1944/9/8/679/s1. Figure S1: Detachment of surface microparticles form a silicone oil droplet. Initially, the stable particle film was formed using *E*-field of 200 Vmm^{-1}, DC. When the polarization of the *E*-field changes and its intensity is increased to more than 800 Vmm^{-1}, the particle film undergoes one cycle of shaking. At the compression stage, the compressed particle film starts to crumple and eventually irreversibly detach (some or all particles depending on the *E*-field strength) from the surface of the droplet. The droplet is imaged parallel to the direction of *E*-field through transparent electrodes. The diameter of the droplet is ~1.7 mm. Figure S2: (a) The sample cell was made of glass (15 mm × 15 mm × 30 mm) where two of the walls are coated with electrically conductive ITO layers. The high-voltage bipolar signal was provided to the cell via two crocodile clips attached to the ITO electrodes. The transparent ITO electrodes allow for observation in a direction along the electric field. A droplet containing colloidal particles is made using a mechanical pipette; (b) Modification of the surface chemistry of the polystyrene particles (PS140) resulted in a change of their contact angle at the castor oil–silicone oil interface, Movie S1: Assembly of surface particles by the EHD convective flows, Movie S2: Electro-shaking particle film, Movie S3: Intruder motion through the particle film, Movie S4: Preliminary experiment on particle size-segregation.

Acknowledgments: Zbigniew Rozynek acknowledges financial support from the National Science Centre, Poland, through the FUGA4 programme (2015/16/S/ST3/00470) and through the OPUS programme (2015/19/B/ST3/03055).

Author Contributions: Zbigniew Rozynek initiated the project and designed all experiments. Agnieszka Magdziarz modified polystyrene particles. Zbigniew Rozynek and Milena Kaczmarek-Klinowska performed the experiments. All authors contributed to the data analysis and the presentation of the results, and the authors took part in discussions towards the finalization of the manuscript. Zbigniew Rozynek authored the paper.

Conflicts of Interest: The authors declare that the research was conducted without any commercial or financial relationships that could be construed as a potential conflict of interest.

References

1. Wang, Y.; Wang, Y.; Breed, D.R.; Manoharan, V.N.; Feng, L.; Hollingsworth, A.D.; Weck, M.; Pine, D.J. Colloids with valence and specific directional bonding. *Nature* **2012**, *491*, 51–55. [CrossRef] [PubMed]
2. Evers, C.H.J.; Luiken, J.A.; Bolhuis, P.G.; Kegel, W.K. Self-assembly of microcapsules via colloidal bond hybridization and anisotropy. *Nature* **2016**, *534*. [CrossRef] [PubMed]
3. Vasilyev, O.A.; Klumov, B.A.; Tkachenko, A.V. Chromatic patchy particles: Effects of specific interactions on liquid structure. *Phys. Rev. E* **2015**, *92*. [CrossRef] [PubMed]
4. Han, M.; Wu, H.; Luijten, E. Electric double layer of anisotropic dielectric colloids under electric fields. *Eur. Phys. J. Spec. Top.* **2016**, *225*, 685–698. [CrossRef]
5. Sander, J.S.; Studart, A.R. Monodisperse Functional Colloidosomes with Tailored Nanoparticle Shells. *Langmuir* **2011**, *27*, 3301–3307. [CrossRef] [PubMed]
6. Song, P.C.; Wang, Y.F.; Wang, Y.; Hollingsworth, A.D.; Weck, M.; Pine, D.J.; Ward, M.D. Patchy Particle Packing under Electric Fields. *J. Am. Chem. Soc.* **2015**, *137*, 3069–3075. [CrossRef] [PubMed]

7. Pethig, R. Review Article-Dielectrophoresis: Status of the theory, technology, and applications. *Biomicrofluidics* **2010**, *4*. [CrossRef]
8. Baraban, L.; Streubel, R.; Makarov, D.; Han, L.; Karnaushenko, D.; Schmidt, O.G.; Cuniberti, G. Fuel-Free Locomotion of Janus Motors: Magnetically Induced Thermophoresis. *ACS Nano* **2013**, *7*, 1360–1367. [CrossRef] [PubMed]
9. Bickel, T.; Zecua, G.; Wuerger, A. Polarization of active Janus particles. *Phys. Rev. E* **2014**, *89*. [CrossRef] [PubMed]
10. Rozynek, Z.; Józefczak, A. Patchy colloidosomes—An emerging class of structures. *Eur. Phys. J. Spec. Top.* **2016**, *225*, 741–756. [CrossRef]
11. Loget, G.; Kuhn, A. Bulk synthesis of Janus objects and asymmetric patchy particles. *J. Mater. Chem.* **2012**, *22*, 15457–15474. [CrossRef]
12. Pawar, A.B.; Kretzschmar, I. Fabrication, assembly, and application of patchy particles. *Macromol. Rapid Commun.* **2010**, *31*, 150–168. [CrossRef] [PubMed]
13. Rodriguez-Fernandez, D.; Liz-Marzan, L.M. Metallic Janus and Patchy Particles. *Part. Part. Syst. Charact.* **2013**, *30*, 46–60. [CrossRef]
14. Shah, A.A.; Schultz, B.; Kohlstedt, K.L.; Glotzer, S.C.; Solomon, M.J. Synthesis, assembly, and image analysis of spheroidal patchy particles. *Langmuir* **2013**, *29*, 4688–4696. [CrossRef] [PubMed]
15. Walther, A.; Muller, A.H. Janus particles: Synthesis, self-assembly, physical properties, and applications. *Chem. Rev.* **2013**, *113*, 5194–5261. [CrossRef] [PubMed]
16. Subramaniam, A.B.; Abkarian, M.; Stone, H.A. Controlled assembly of jammed colloidal shells on fluid droplets. *Nat. Mater.* **2005**, *4*, 553–556. [CrossRef] [PubMed]
17. Rozynek, Z.; Castberg, R.; Kalicka, A.; Jankowski, P.; Garstecki, P. Electric field manipulation of particles in leaky dielectric liquids. *Arch. Mech.* **2015**, *67*, 385–399.
18. Rozynek, Z.; Mikkelsen, A.; Dommersnes, P.; Fossum, J.O. Electroformation of Janus and patchy capsules. *Nat. Commun.* **2014**, *5*. [CrossRef] [PubMed]
19. Dommersnes, P.; Rozynek, Z.; Mikkelsen, A.; Castberg, R.; Kjerstad, K.; Hersvik, K.; Otto Fossum, J. Active structuring of colloidal armour on liquid drops. *Nat. Commun.* **2013**, *4*. [CrossRef] [PubMed]
20. Nudurupati, S.; Janjua, M.; Aubry, N.; Singh, P. Concentrating particles on drop surfaces using external electric fields. *Electrophoresis* **2008**, *29*, 1164–1172. [CrossRef] [PubMed]
21. Li, M.; Li, D. Redistribution of charged aluminum nanoparticles on oil droplets in water in response to applied electrical field. *J. Nanopart. Res.* **2016**, *18*. [CrossRef]
22. Nudurupati, S.; Janjua, M.; Singh, P.; Aubry, N. Effect of parameters on redistribution and removal of particles from drop surfaces. *Soft Matter* **2010**, *6*, 1157–1169. [CrossRef]
23. Amah, E.; Shah, K.; Fischer, I.; Singh, P. Electrohydrodynamic manipulation of particles adsorbed on the surface of a drop. *Soft Matter* **2016**, *12*, 1663–1673. [CrossRef] [PubMed]
24. Rozynek, Z.; Dommersnes, P.; Mikkelsen, A.; Michels, L.; Fossum, J.O. Electrohydrodynamic controlled assembly and fracturing of thin colloidal particle films confined at drop interfaces. *Eur. Phys. J. Spec. Top.* **2014**, *223*, 1859–1867. [CrossRef]
25. Rozynek, Z.; Zacher, T.; Janek, M.; Caplovicova, M.; Fossum, J.O. Electric-field-induced structuring and rheological properties of kaolinite and halloysite. *Appl. Clay Sci.* **2013**, *77–78*, 1–9. [CrossRef]
26. Rozynek, Z.; Wang, B.; Fossum, J.O.; Knudsen, K.D. Dipolar structuring of organically modified fluorohectorite clay particles. *Eur. Phys. J. E Soft Matter* **2012**, *35*. [CrossRef] [PubMed]
27. Uñac, R.O.; Benito, J.G.; Vidales, A.M.; Pugnaloni, L.A. Arching during the segregation of two-dimensional tapped granular systems: Mixtures versus intruders. *Eur. Phys. J. E* **2014**, *37*, 1–9. [CrossRef] [PubMed]
28. Yang, X.Q.; Zhou, K.; Qiu, K.; Zhao, Y.M. Segregation of large granules from close-packed cluster of small granules due to buoyancy. *Phys. Rev. E* **2006**, *73*. [CrossRef] [PubMed]
29. Tai, C.H.; Hsiau, S.S.; Kruelle, C.A. Density segregation in a vertically vibrated granular bed. *Powder Tech.* **2010**, *204*, 255–262. [CrossRef]
30. Goetzendorfer, A.; Tai, C.-H.; Kruelle, C.A.; Rehberg, I.; Hsiau, S.-S. Fluidization of a vertically vibrated two-dimensional hard sphere packing: A granular meltdown. *Phys. Rev. E* **2006**, *74*. [CrossRef] [PubMed]

materials

MDPI

Article

Effect of Geometric and Chemical Anisotropy of Janus Ellipsoids on Janus Boundary Mismatch at the Fluid–Fluid Interface

Dong Woo Kang †, Woong Ko †, Bomsock Lee and Bum Jun Park *

Department of Chemical Engineering, Kyung Hee University, Yongin, Gyeonggi-do 17104, Korea;
lukekang070@gmail.com (D.W.K.); wng55555@gmail.com (W.K.); bslee@khu.ac.kr (B.L.)
* Correspondence: bjpark@khu.ac.kr; Tel.: +82-31-201-2429
† These authors contributed equally to this work.

Academic Editor: To Ngai
Received: 20 June 2016; Accepted: 4 August 2016; Published: 6 August 2016

Abstract: We investigated the geometric and chemical factors of nonspherical Janus particles (i.e., Janus ellipsoids) with regard to the pinning and unpinning behaviors of the Janus boundary at the oil–water interface using attachment energy numerical calculations. The geometric factors were characterized by aspect ratio (AR) and location of the Janus boundary (α) separating the polar and apolar regions of the particle. The chemical factor indicated the supplementary wettability (β) of the two sides of the particle with identical deviations of apolarity and polarity from neutral wetting. These two factors competed with each other to determine particle configurations at the interface. In general, the critical value of β (β_c) required to preserve the pinned configuration was inversely proportional to the values of α and AR. From the numerical calculations, the empirical relationship of the parameter values of Janus ellipsoids was found; that is, $\lambda = \Delta\beta_c/\Delta\alpha \approx 0.61 AR - 1.61$. Particularly for the Janus ellipsoids with $AR > 1$, the β_c value is consistent with the boundary between the tilted only and the tilted equilibrium/upright metastable region in their configuration phase diagram. We believe that this work performed at the single particle level offers a fundamental understanding of the manipulation of interparticle interactions and control of the rheological properties of particle-laden interfaces when particles are used as solid surfactants.

Keywords: Janus particle; fluid–fluid interface; attachment energy; configuration; Janus boundary

1. Introduction

Typical colloidal particles with appropriate wettability can be irreversibly adsorbed at fluid–fluid interfaces (e.g., oil–water and air–water interfaces) [1,2]. Strong particle adsorptions can decrease interfacial tension and stabilize the interface, thereby preventing phase separation and coalescence in emulsion systems. In particular, chemically homogeneous colloidal particles likely impart kinetic stability in Pickering emulsions stabilized by solid particles [3,4]. The efficacy of interfacial stability is highly dependent on the relative amount of each fluid phase and particle wettability (polarity or apolarity) with regard to the fluid–fluid interface that can be characterized via three-phase contact angles.

Janus particles that possess chemical and/or geometric anisotropy can improve the stabilization efficiency of Pickering emulsion systems when used as stabilizers [5–15]. This is because Janus particles tend to be aligned toward increasing the surface area of preferred wetting (e.g., apolar surfaces exposed to oil and polar surfaces exposed to water), resulting in effectively reduced surface free energy. Additional degrees of freedom gained from anisotropic properties can lead to diversity in configurational behaviors (upright or tilted) when attached to a fluid–fluid interface [16–20]. It is important to note that configurations on the single-particle level can significantly affect interactions between particles at the interface, their assembly structures, and consequently the rheological properties of the interface [21–31]. For instance, Janus spheres with two chemically different sides (apolar and polar surfaces) generally adopt an upright orientation corresponding to the geometry in which the Janus boundary or wettability separating line (WSL) is pinned to the fluid–fluid interface [32]. With regard to non-spherical Janus particles (e.g., Janus ellipsoids, Janus dumbbells, and Janus cylinders), particle configurations can be either upright or tilted, depending on the relative influence of chemical and geometric factors. Particularly, particles adopted tilted configurations when the geometric effect (e.g., aspect ratio, *AR*) is stronger than the chemical effect (e.g., wettability), whereas particles under the opposite conditions adopted upright configurations. Janus particles with tilted configurations likely deformed their surrounding fluid interface due to unpreferred wetting (e.g., apolar surfaces exposed to water, and vice versa), and the resulting interface deformation led to lateral capillary interactions between particles to minimize the surface areas of the deformed interfaces [33–39]. Capillary interactions between particles at the interface directly affected their microstructure, and therefore the rheological properties of the particle-laden interface. Effects of configuration at the single-particle level on interparticle interactions were experimentally demonstrated using Janus cylinders with apolar and polar surfaces prepared via PDMS (Polydimethylsiloxane) molding techniques [33,34,40]. Apolar and polar precursor solutions were added to a PDMS mold with cylindrical wells one after another, and subsequent UV exposure led to photopolymerization of the precursor solution, resulting in Janus cylinders. The aspect ratio and relative lengths of the two sides could be readily controlled depending on the well depth in the mold and the amount of each precursor solution. Janus cylinders with high *AR* values adopted tilted configurations and exhibited strong capillary attractions. In contrast, low *AR* particles adopted upright configurations and attractive interactions occurred negligibly.

The configurational behaviors of two types of non-spherical Janus particles (i.e., Janus ellipsoids and Janus dumbbells) have been theoretically investigated by numerically calculating the attachment energy at the fluid–fluid interface as a function of orientation angle [16,17]. With regard to Janus dumbbells, a single energy minimum occurred in the attachment energy profile, indicating that the Janus dumbbells exclusively adopted either an upright or a tilted configuration. Interestingly, for Janus ellipsoids with particular *AR* and wettability values, two energy minima appeared in the attachment energy profile; one corresponding to an upright configuration, and the other to a tilted configuration—one of the two configurations should be kinetically stable. With regard to Janus ellipsoids, it is unknown whether the Janus boundary of the particles can be detached or unpinned from the fluid–fluid interface, regardless of configuration. These pinning and unpinning behaviors might depend on the relative effect of chemical and geometric anisotropy. Indeed, it was previously reported that the Janus boundary of Janus spheres could be unpinned from the interface when the boundary significantly deviated from the central region of the particles [2,32]. In this study, we quantitatively investigate the effects of geometric and chemical factors of Janus ellipsoids on the pinning and unpinning behaviors of the Janus boundary at the fluid interface. We also relate the unpinned configuration to the tilted configuration when the *AR* value is larger than 1.

2. Theoretical Basis

To characterize the geometric and chemical anisotropy of Janus ellipsoids, we define the aspect ratio (*AR*), the location of the Janus boundary (α), and the supplementary wettability (β) as shown in Figure 1a,b. The aspect ratio is the ratio of the major axis to the minor axis, calculated by $AR = c/a$. Depending on the value of *AR*, the ellipsoids are prolate or oblate at $AR > 1$ and $AR < 1$, respectively, with a sphere formed at $AR = 1$. We designate all studied particles as Janus ellipsoids. The Janus boundary that indicates the wettability separation line (WSL) dividing polar and apolar regions is characterized by an elevation angle, α. The supplementary wettability is defined as $\beta = 90° - \theta_P = \theta_A - 90°$, in which θ_P and θ_A indicate the three-phase contact angles of homogeneous polar ($\alpha = 180°$) and apolar ($\alpha = 0°$) spheres when trapped at an oil–water interface, respectively (Figure 1b).

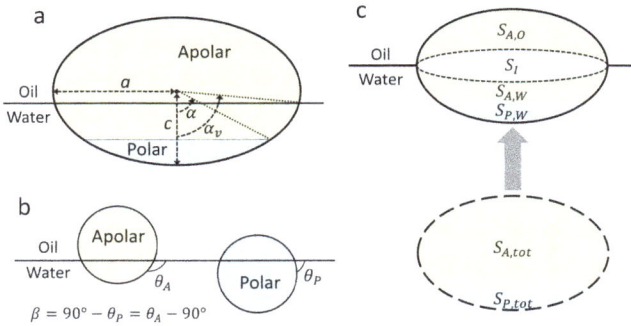

Figure 1. Schematics of a Janus ellipsoid for attachment energy calculations. (**a**) Geometric relationships of the Janus ellipsoid; (**b**) Three-phase contact angles of homogeneous polar and apolar spheres at the oil–water interface; (**c**) Schematic illustration of Janus ellipsoid attachment to the oil–water interface.

Based on these geometric relationships, the attachment energy of a particle that is attached to a planar oil–water interface from the water phase (Figure 1c) is given by [1,2,16,17]

$$\Delta E_{Iw} = E_I - E_w \tag{1}$$

The respective energies of a particle at the oil–water interface (E_I) and a particle immersed in the aqueous phase (E_w) are

$$E_I = S_{Ao}\gamma_{Ao} + S_{Aw}\gamma_{Aw} + S_{Po}\gamma_{Po} + S_{Pw}\gamma_{Pw} + S_I^{(2)}\gamma_{ow} \tag{2}$$

$$E_w = S_A^{tot}\gamma_{Aw} + S_P^{tot}\gamma_{Pw} + S_I^{(1)}\gamma_{ow} \tag{3}$$

where S_{ij} is the surface area ($i = P$ (polar) or A (apolar)) exposed to a fluid phase ($j = w$ (water) or o (oil)), γ_{ij} is the corresponding surface energy of surface i and fluid j, $S_I^{(2)}$ and $S_I^{(1)}$ are the surface area values of the oil–water interface when particles are present and absent, respectively, and $S_I = S_I^{(2)} - S_I^{(1)}$ indicates the displaced area of the interface due to presence of the particle. The displaced area (S_I) corresponds to the cross-sectional area of the particle at α_v, which is the elevation angle measured from the major axis to the fluid interface (i.e., the three phase contact line), as shown in Figure 1a. S_A^{tot} and S_P^{tot} are the total surface area values of the apolar and polar regions of the particle, respectively. Substitution of the following Young's Equations:

$$\gamma_{ow}\cos\theta_P = \gamma_{Po} - \gamma_{Pw} \text{ for the polar surface} \tag{4}$$

$$\gamma_{ow}\cos\theta_A = \gamma_{Ao} - \gamma_{Aw} \text{ for the apolar surface} \tag{5}$$

into Equations (1)–(3) yields the following:

$$\Delta E_{Iw} = \gamma_{ow}\left(S_{Ao}\cos\theta_A + S_{Po}\cos\theta_P - S_I\right) \text{ from the water phase} \tag{6}$$

Similarly, the attachment energy of a Janus particle from the oil phase to the oil–water interface can be expressed as:

$$\Delta E_{Io} = -\gamma_{ow}\left(S_{Aw}\cos\theta_A + S_{Pw}\cos\theta_P + S_I\right) \text{ from the oil phase.} \tag{7}$$

The difference between Equations (6) and (7) is the reference energy state (i.e., E_w and E_o); therefore, the shapes of the attachment energy profiles obtained from the equations should be identical.

Since the use of either Equation (6) or (7) predicts the same configurational behavior, Equation (6) is used in this work. To calculate the surface area S_{ij}, a numerical method using the hit-and-miss Monte Carlo method is employed [16]. The oil–water interfacial tension value is arbitrarily designated as γ_{ow} = 50 mN/m, corresponding to the interfacial tension of a decane–water interface.

In the numerical calculations, it is assumed that the meniscus at the particle interface is smooth. The effect of line tension at the three-phase contact line can be neglected when the particle dimensions are sufficiently large (e.g., >10 nm) [15,41]. In this case, the effect of the Brownian motion of particles on the configuration behaviors is also negligible [16]. The oil–water interface around the particle is assumed to be flat when the particle size is less than a few hundred microns, in which the corresponding Bond number (ratio of gravitational force to surface tension force) is sufficiently small. Although interfacial deformation around the particles is possible for non-spherical Janus particles, it was previously demonstrated that the flat interface assumption (FIA) with regard to the attachment energy calculations was appropriate to determine the equilibrium configurations of the particles at the fluid–fluid interface [16]. Moreover, it was reported that the configuration behaviors of Janus cylinders predicted from the numerical calculations based on the FIA showed a good agreement with the experimental observations [33,34]. Note that in spite of the consistency between the experimental and theoretical results for the configuration behaviors when the FIA is used, interfacial deformation around Janus particles with a tilted configuration can occur, leading to capillary-induced interactions between the tilted Janus particles [33–35]. Based on these conditions and methods, the attachment energy is calculated as a function of vertical position, α_v. To evaluate whether the Janus boundary of the particles is pinned or unpinned at the oil–water interface, the attachment energy is minimized to obtain the lowest energy minimum, $\Delta E_{att}^{min}(\alpha_v)$, and the corresponding vertical position , $\alpha_{v,min}$.

Notably, for Janus prolates with $AR > 1$, there are four possible cases in their configuration behaviors, depending on AR and wettability values, such as the upright only, the upright equilibrium/tilted metastable, the tilted equilibrium/upright metastable, and the tilted only regions [16,17]. For the co-existing regions, for instance, two energy minima appear in the attachment energy profiles, indicating that particles can adopt either upright or tilted configurations, whereas particles in the titled only region possess only the tilted configuration. Therefore, to further determine the configuration behaviors (upright or/and tilted) of the Janus prolates, the attachment energy is calculated as functions of the vertical position (α_v) as well as the orientation angle ($0° < \theta_r < 90°$), in which $\theta_r = 0°$ and $90°$ indicate the geometries of the Janus boundary parallel and perpendicular to the interface, respectively. In this case, the attachment energy is scanned by varying the α_v values at a constant θ_r. The minimum attachment energy is then calculated at the given θ_r value, and the same procedure is repeated for different values of θ_r to find a global energy minimum, $\Delta E_{att}^{min}(\alpha_v, \theta_r)$.

For spherical Janus particles with supplementary wettability (i.e., $\cos\theta_A = -\sin\beta$ and $\cos\theta_A = \sin\beta$), the attachment energy from the water phase (Equation (6)) can be further simplified using geometric relationships. In particular, when the Janus boundary of a Janus sphere is pinned at the oil–water interface, the surface areas of the apolar and polar regions exposed to the oil phase are $S_{Ao} = 2\pi a^2(1 + \cos\alpha)$ and $S_{Po} = 0$, respectively. The displaced area in this geometry is $S_I = \pi a^2 \sin^2\alpha$. Therefore, Equation (6) becomes,

$$\Delta E_{Iw} = -\pi a^2 \gamma_{ow}\left(2(1 + \cos\alpha)\sin\beta + \sin^2\alpha\right) \text{ from the water phase} \tag{8}$$

Equation (8) should satisfy the condition of $\frac{d(\Delta E_{Iw})}{d\alpha} = 0$ at $\beta = \beta_c$ for the Janus sphere with the pinned Janus boundary, yielding $\sin\beta_c = \cos\alpha$, and thus $\beta_c = 90° - \alpha$.

3. Results and Discussion

Pinning ($\alpha = \alpha_{v,min}$) and unpinning ($\alpha \neq \alpha_{v,min}$) behaviors of the Janus particles with respect to the interfaces can be determined through two competing factors: geometric and chemical anisotropy. To minimize the attachment energy in Equation (6), a Janus particle tends to be aligned toward increasing the displaced area (S_I) at the oil–water interface, simultaneously increasing the surface area of the preferred wetting state (S_{Ao}; note that $\cos\theta_A < 0$). The state of preferred wetting indicates the configuration of the apolar and polar regions exposed to their favorable fluid phases, oil and water, respectively. Depending on the relative influence of these two factors, the particle is likely to preferentially adopt the pinned configuration when wettability effects are greater than geometric effects, whereas particles are unpinned from the interface when geometric effects are dominant. Notably, Janus prolates can adopt the two configurations, the upright and tilted one (e.g., when the *AR* value is high and wettability is moderate) [16]. In this particular case, the attachment energy may be further decreased by rotating the particle at the interface. We also examine the relationship between the pinning/unpinning and the upright/tilted configuration behaviors.

To investigate the effects of geometry (*AR* and α) and wettability (β) on the pinning and unpinning behaviors of Janus ellipsoids, the attachment energy (ΔE_{Iw}) is calculated as a function of α_v at constant values of β and *AR* = 0.5 (oblate), 1 (sphere), or 2 (prolate). Since the supplementary wettability (β) is used and Janus particles with $\alpha \leqslant 90°$ are considered, ΔE_{Iw} is calculated in the range of $0° \leqslant \alpha_v \leqslant 90°$. Note that similar results can be obtained in the case of Janus ellipsoids with $\alpha > 90°$.

For the Janus ellipsoids with $\alpha = 90°$—in which the polar and apolar surface area values are identical—the lowest energy minimum (ΔE_{att}^{min}) for all cases is found at $\alpha_{v,min} = 90°$, as shown in Figure 2a–c. The result of $\alpha = \alpha_{v,min}$ indicates that the Janus boundary of the geometrically symmetric Janus ellipsoids is always pinned at the oil–water interface, regardless of the values of *AR* and β. The pinned configuration of the particles simultaneously satisfies both conditions; that is, the particles tend to possess maximum values of the displaced area (S_I) and preferred wetting (S_{Ao}) in Equation (6) when $\alpha = \alpha_{v,min}$.

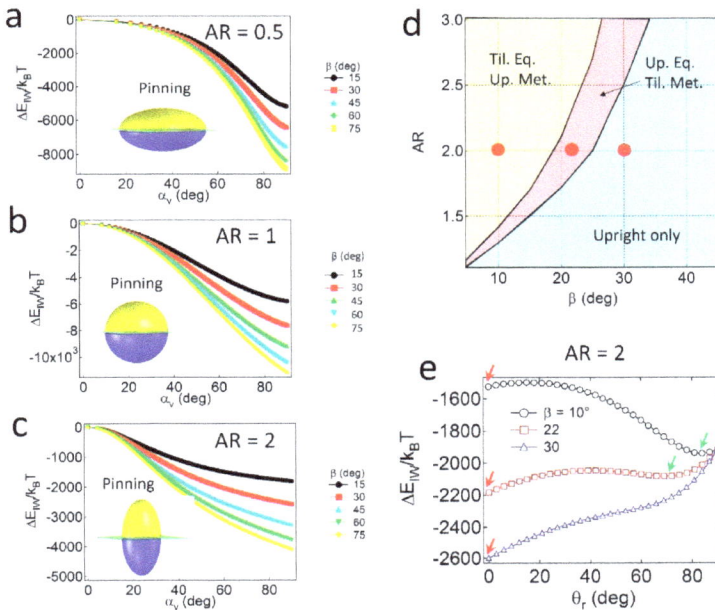

Figure 2. Attachment energy profiles of Janus ellipsoids with $\alpha = 90°$ and (**a**) Aspect ratio (*AR*) = 0.5 (oblate); (**b**) 1 (sphere); or (**c**) 2 (prolate). In all cases, the Janus boundary is pinned at the oil–water interface; (**d**) Configuration phase diagram of Janus prolates as functions of the *AR* and β values; (**e**) The attachment energy profiles of Janus prolates as a function of the orientation angle (θ_r).

Janus prolates with particular AR and β values can also adopt the tilted configuration. As shown in the configuration phase diagram in Figure 2d, particles with relatively high β and low AR values only possess the upright or the pinned configuration (light blue region), whereas two coexisting regions (pink and yellow regions) appear when particles carry the opposite properties (i.e., relatively low β and high AR values). For instance, for particles with $AR = 2$ and $\beta = 10°$ or $22°$ (denoted as red dots in Figure 2d), two energy minima exist in the attachment energy profiles in Figure 2e; that is, one corresponds to the upright (red arrow) and the other to the tilted one (green arrow). In contrast, a single energy minimum for the particle with $AR = 2$ and $\beta = 30°$ (Figure 2e) indicates that the particle solely adopts the upright configuration. Notably, the absence of a tilted only region in Figure 2d might be related to the absence of the unpinned configuration. In other words, since the Janus boundary of the particles with $\alpha = 90°$ can always be pinned to the interface, regardless of the values of AR and β, at least one of the energy minima in the attachment energy profiles should exist at $\theta_r = 0$.

When the particles are geometrically asymmetric ($\alpha = 60°$), configuration behavior depends on the relative geometry and wettability effects. As shown in Figure 3a, for oblates with $AR = 0.5$, the value of $\alpha_{v,min}$ at ΔE_{att}^{min} gradually decreases with β, and is not consistent with α if the value of β is not sufficiently high. The condition of $\alpha_{v,min} = \alpha$ indicates that a pinned geometry is found at a critical value of β, in which $\beta_c = 67°$. Similar results are found for spheres with $AR = 1$ (Figure 3b) and prolates with $AR = 2$ (Figure 3c) when $\alpha = 60°$, in which the critical wettability values are $\beta_c = 30°$ and $8°$, respectively. The increase in β_c for lower AR particles is due to the relatively large displaced area (S_I) around the central regions of the particles; thus, the geometric effects become stronger than the wettability effects. Note that the value of $\beta_c = 30°$ for the Janus sphere obtained from the numerical calculation is consistent with the theoretical prediction of $\beta_c = 90° - \alpha$, validating the numerical method.

The configuration behaviors of geometrically asymmetric Janus prolates are also affected by the presence of the secondary energy minimum in the attachment energy profile. As shown in the configuration phase diagram of Janus prolates with $\alpha = 60°$ in Figure 3d, four distinct regions are observed; the tilted only (light green), the tilted equilibrium/upright metastable (yellow), the upright equilibrium/tilted metastable (pink), and the upright only (light blue) regions. The example attachment energy profiles as a function of θ_r are shown in Figure 3e, and the corresponding energy minima are indicated by red and green arrows. Unlike the Janus prolates with $\alpha = 90°$, it is interesting to note that the tilted only region is found at the conditions of relatively high AR and low β values. The presence of the tilted only region is likely due to the presence of the unpinning configuration of the particles; that is, when the prolate particles are unpinned, no energy minimum exists at $\theta_r = 0$, and consequently, the particles only adopt the tilted configuration. In contrast, when the Janus boundary of prolate particles are pinned to the interface in Figure 3c, they essentially possess an energy minimum at $\theta_r = 0$, leading to the upright/tilted coexisting regions (yellow and pink areas in Figure 3d) or the upright only region (light blue in Figure 3d). Note that the $\beta_c = 8°$ value for the Janus prolate with $AR = 2$ in Figure 3c shows a good agreement with the boundary between the light green and yellow regions in the Figure 3d.

For Janus ellipsoids with a high degree of geometric asymmetry ($\alpha = 30°$), strong wettability effects are required to preserve the pinned configuration. As shown in Figure 4a, for oblates with $AR = 0.5$, the value of $\alpha_{v,min}$ consistently decreases as β increases, and does not match the value of $\alpha = 30°$ up to $\beta = 50°$. Due to large values of S_I around the central regions of the oblate, geometric effects are likely to be dominant to wettability effects, resulting in particles with central regions located at the interface, adopting an unpinned configuration. For Janus ellipsoids with lower AR values, wettability effects gradually become significant. As shown in Figure 4b,c, the Janus boundaries of Janus ellipsoids with $AR = 1$ and 2 are pinned to the interface when the critical wettability value is $\beta_c = 60°$ or $22°$. Note that the value of β_c increases as the particles become more geometrically asymmetric ($\beta_c = 30°$ and $8°$ for particles ($\alpha = 60°$) with $AR = 1$ and 2, respectively).

Similar to the case of the Janus prolates with $\alpha = 60°$, the four distinct regions in the configuration phase diagram for the prolate particles with $\alpha = 30°$ are found, as shown in Figure 4d. Example attachment energy profiles as a function of θ_r, representing each configuration region and the location of energy minima, are shown in Figure 4e. The tilted only region (light green area) in relatively high AR and low β values corresponds to the condition where the Janus boundary of particles is unpinned. For particles with $AR = 2$, for instance, the boundary between the light green and the yellow regions in Figure 4e is consistent with the critical wettability value, $\beta_c = 22°$ in Figure 4c. Except for the tilted only region, a portion of the particles always adopts the upright configuration due to the presence of an energy minimum at the pinned geometry in Figure 4c.

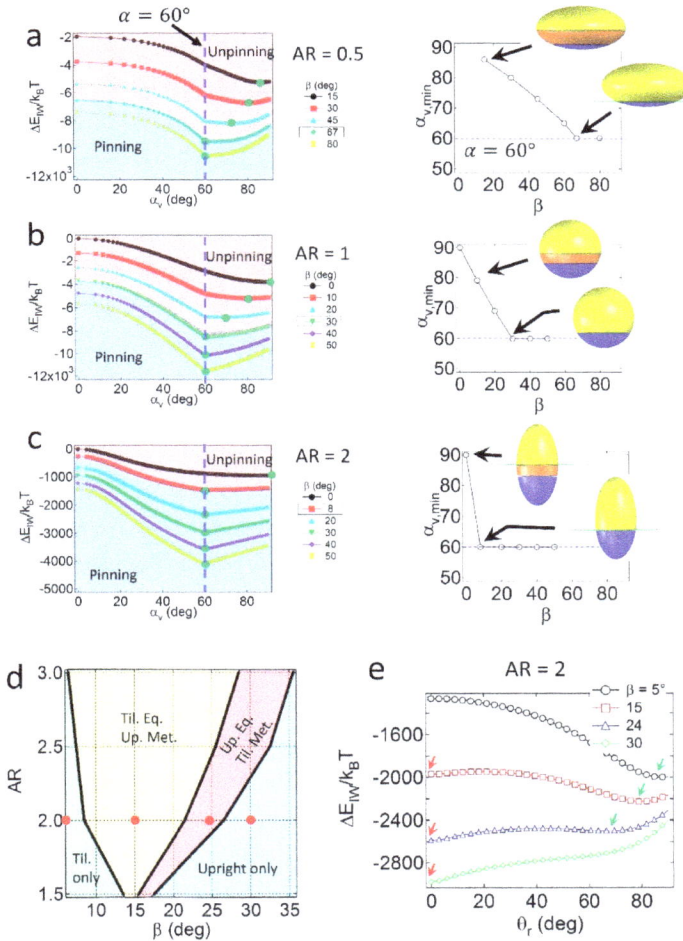

Figure 3. Attachment energy profiles of Janus particles with $\alpha = 60°$ (vertical dashed line) and (a) $AR = 0.5$; (b) 1; or (c) 2. Pink and blue regions in each plot represent unpinned and pinned configurations, respectively. Green circles denote the lowest energy minimum (ΔE_{att}^{min}) and the corresponding value of α ($\alpha_{v,min}$). The effect of β on $\alpha_{v,min}$ is shown on the right; (d) Configuration phase diagram of Janus prolates as functions of the AR and β values; (e) The attachment energy profiles of Janus prolates as a function of the orientation angle (θ_r).

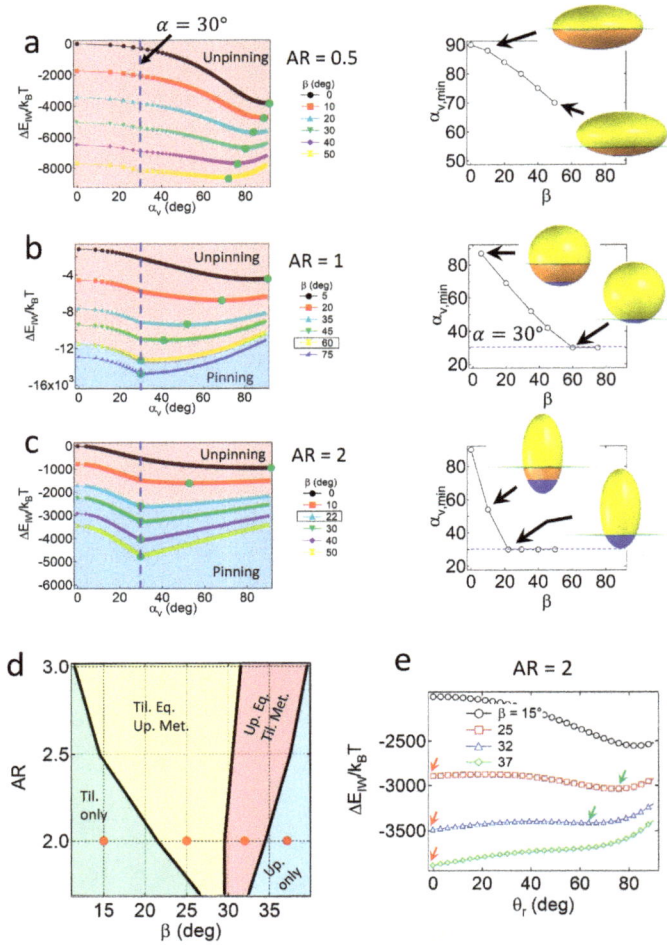

Figure 4. Attachment energy profiles of Janus particles with $\alpha = 30°$ (vertical dashed line) and (**a**) $AR = 0.5$; (**b**) 1; or (**c**) 2. Green circles denote the values of ΔE_{att}^{min} and $\alpha_{v,min}$. The relationship between β and $\alpha_{v,min}$ is shown on the right; (**d**) Configuration phase diagram of Janus prolates as functions of the AR and β values; (**e**) The attachment energy profiles of Janus prolates as a function of the orientation angle (θ_r).

Additionally, the critical values of β are further calculated as functions of AR and α. As shown in Figure 5a, the regions above each curve correspond to the pinned Janus boundary at the interface, whereas the regions below the curve indicate that the Janus boundary is unpinned from the interface. In general, the value of β_c (due to wettability effects) is inversely proportional to the values of α and AR, which are due to geometric effects. As the values of AR decrease, the particles demonstrate reduced attachment energy when the interface is located at the central regions of the particles. When the value of α decreases (the particles become more geometrically asymmetric), a higher value of β is required to adopt a pinned configuration. More quantitatively, to find a relationship of β_c, α, and AR, the β_c values are plotted as a function of α, as shown in Figure 5b. Then, slopes ($\lambda = \Delta\beta_c/\Delta\alpha$) obtained from linear regression of each line representing different AR values are shown in the inset plot in Figure 5b, in which the slope ($\Delta\lambda/\Delta AR$) is found to be ~0.61 ± 0.06. Based on the values of $\lambda = -1$ when $AR = 1$, an empirical relationship for Janus ellipsoids is obtained, $\lambda = \Delta\beta_c/\Delta\alpha \approx 0.61 AR - 1.61$.

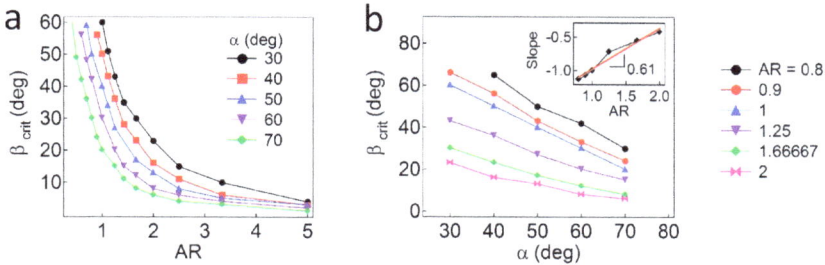

Figure 5. Critical wettability values (β_c) as functions of AR and α. (**a**) Plot of β_c versus AR. The regions above and below each curve correspond to the pinned and unpinned configurations, respectively; (**b**) Plot of β_c versus α. The inset indicates the slope of each line in the plot (β_c versus α) as a function of AR.

Finally, it is notable that when the Janus boundary of Janus prolates ($AR > 1$) with $\beta < \beta_c$ is unpinned, they rotate and adopt the tilted configuration to further decrease their attachment energy by increasing the displaced area (S_l) [16,17] and, consequently, all portions of the particles would adopt the tilted configuration. In this case, the β_c value corresponds to the boundary between two configuration regions—the tilted only and the tilted equilibrium/upright metastable regions.

4. Conclusions

Pinning and unpinning behaviors of the Janus boundaries of Janus ellipsoids at the oil–water interface were quantitatively investigated using numerical calculations of the attachment energy. Particle configurations were determined via two competing factors: geometric (α and AR) and chemical (β) anisotropy values. For geometrically and chemically symmetric Janus ellipsoids ($\alpha = 90°$), the Janus boundary was always pinned to the interface, regardless of the values of AR and β. Contrarily, for geometrically asymmetric Janus ellipsoids with $\alpha \neq 90°$, the Janus boundary was unpinned when chemical effects (β) were not sufficiently high. In general, particles with large values of α and AR required stronger wettability to preserve the Janus boundary pinned to the interface. It was also found that determination of the critical values of β as functions of α and AR yielded the empirical relationship $\lambda = \Delta\beta_c/\Delta\alpha \approx 0.61 AR - 1.61$. It was notable that for the Janus prolates, the value of β_c corresponded to the boundary between the tilted only and the tilted equilibrium/upright metastable regions in their configuration phase diagram. In the case of non-supplementary wettability where $90° - \theta_P \neq \theta_A - 90°$, the pinning and unpinning behaviors are likely similar to the case of supplementary wetting when the orientation angle is not considered ($AR \leqslant 1$). For Janus prolates that can adopt tilted configurations, it was reported that the relative strength of θ_P and θ_A significantly affected the tilted angle and the configuration behaviors [17]. We believe that this work offers fundamental ideas with regard to the attachment and configurational behaviors of nonspherical Janus particles that consequently enable manipulation of interparticle interactions and control of rheological properties of interfaces when used as solid surfactants.

Acknowledgments: This work was supported by the Engineering Research Center of Excellence Program of the Korea Ministry of Science, ICT & Future Planning (MSIP)/National Research Foundation of Korea (NRF-2014R1A5A1009799).

Author Contributions: Dong Woo Kang and Woong Ko contributed equally to this work. They performed the numerical calculations and contributed in the discussion. Bomsock Lee and Bum Jun Park planned the research, performed the calculations, and wrote the manuscript.

Conflicts of Interest: The authors declare no conflict of interest.

References

1. Binks, B.P.; Horozov, T.S. *Colloidal Particles at Liquid Interfaces*; Cambridge University Press: New York, NY, USA, 2006.
2. Park, B.J.; Lee, D.; Furst, E.M. Chapter 2 interactions and conformations of particles at fluid–fluid interfaces. In *Particle-Stabilized Emulsions and Colloids: Formation and Applications*; The Royal Society of Chemistry: Cambridge, UK, 2015; pp. 8–44.
3. Pickering, S.U. Emulsions. *J. Chem. Soc. Trans.* **1907**, *91*, 2001–2021. [CrossRef]
4. Ramsden, W. Separation of solids in the surface-layers of solutions and 'suspensions' (observations on surface-membranes, bubbles, emulsions, and mechanical coagulation)—Preliminary account. *Proc. R. Soc. Lond.* **1903**, *72*, 156–164. [CrossRef]
5. Casagrande, C.; Fabre, P.; Raphael, E.; Veyssié, M. "Janus beads": Realization and behaviour at water/oil interfaces. *Europhys. Lett.* **1989**, *9*, 251–255. [CrossRef]
6. Casagrande, C.; Veyssié, M. "Grains janus": Réalisation et premières observations des propriétiés interfaciales. *C. R. Acad. Sci. Paris* **1988**, *306*, 1423–1425.
7. Glaser, N.; Adams, D.J.; Boker, A.; Krausch, G. Janus particles at liquid-liquid interfaces. *Langmuir* **2006**, *22*, 5227–5229. [CrossRef] [PubMed]
8. Hu, J.; Zhou, S.; Sun, Y.; Fang, X.; Wu, L. Fabrication, properties and applications of Janus particles. *Chem. Soc. Rev.* **2012**, *41*, 4356–4378. [CrossRef] [PubMed]
9. Kim, J.W.; Cho, J.; Cho, J.; Park, B.J.; Kim, Y.-J.; Choi, K.-H.; Kim, J.W. Synthesis of monodisperse bi-compartmentalized amphiphilic janus microparticles for tailored assembly at the oil–water interface. *Angew. Chem. Int. Ed.* **2016**, *55*, 4509–4513. [CrossRef] [PubMed]
10. Aveyard, R. Can janus particles give thermodynamically stable pickering emulsions? *Soft Matter* **2012**, *8*, 5233–5240. [CrossRef]
11. Tu, F.; Park, B.J.; Lee, D. Thermodynamically stable emulsions using janus dumbbells as colloid surfactants. *Langmuir* **2013**, *29*, 12679–12687. [CrossRef] [PubMed]
12. Walther, A.; Muller, A.H.E. Janus particles. *Soft Matter* **2008**, *4*, 663–668. [CrossRef]
13. Liu, B.; Wei, W.; Qu, X.; Yang, Z. Janus colloids formed by biphasic grafting at a pickering emulsion interface. *Angew. Chem.* **2008**, *120*, 4037–4039. [CrossRef]
14. Kumar, A.; Park, B.J.; Tu, F.; Lee, D. Amphiphilic janus particles at fluid interfaces. *Soft Matter* **2013**, *9*, 6604–6617. [CrossRef]
15. Binks, B.P.; Fletcher, P.D.I. Particles adsorbed at the oil–water interface: A theoretical comparison between spheres of uniform wettability and janus" particles. *Langmuir* **2001**, *17*, 4708–4710. [CrossRef]
16. Park, B.J.; Lee, D. Equilibrium orientation of nonspherical janus particles at fluid–fluid interfaces. *ACS Nano* **2012**, *6*, 782–790. [CrossRef] [PubMed]
17. Park, B.J.; Lee, D. Configuration of nonspherical amphiphilic particles at a fluid–fluid interface. *Soft Matter* **2012**, *8*, 7690–7698. [CrossRef]
18. Lee, C.; Lee, W.B.; Kang, T. Orientation and position of cylindrical-shaped gold nanoparticles at liquid-liquid interfaces. *Appl. Phys. Lett.* **2013**, *103*. [CrossRef]
19. Lewandowski, E.P.; Cavallaro, M.; Botto, L.; Bernate, J.C.; Garbin, V.; Stebe, K.J. Orientation and self-assembly of cylindrical particles by anisotropic capillary interactions. *Langmuir* **2010**, *26*, 15142–15154. [CrossRef] [PubMed]
20. Isa, L.; Samudrala, N.; Dufresne, E.R. Adsorption of sub-micron amphiphilic dumbbells to fluid interfaces. *Langmuir* **2014**, *30*, 5057–5063. [CrossRef] [PubMed]
21. Loudet, J.C.; Alsayed, A.M.; Zhang, J.; Yodh, A.G. Capillary interactions between anisotropic colloidal particles. *Phys. Rev. Lett.* **2005**, *94*. [CrossRef] [PubMed]
22. Rezvantalab, H.; Shojaei-Zadeh, S. Capillary interactions between spherical janus particles at liquid–fluid interfaces. *Soft Matter* **2013**, *9*, 3640–3650. [CrossRef]
23. Wang, J.-Y.; Wang, Y.; Sheiko, S.S.; Betts, D.E.; DeSimone, J.M. Tuning multiphase amphiphilic rods to direct self-assembly. *J. Am. Chem. Soc.* **2011**, *134*, 5801–5806. [CrossRef] [PubMed]

24. Zhang, Z.; Pfleiderer, P.; Schofield, A.B.; Clasen, C.; Vermant, J. Synthesis and directed self-assembly of patterned anisometric polymeric particles. *J. Am. Chem. Soc.* **2010**, *133*, 392–395. [CrossRef] [PubMed]

25. Ruhland, T.M.; Gröschel, A.H.; Walther, A.; Müller, A.H.E. Janus cylinders at liquid-liquid interfaces. *Langmuir* **2011**, *27*, 9807–9814. [CrossRef] [PubMed]

26. Madivala, B.; Fransaer, J.; Vermant, J. Self-assembly and rheology of ellipsoidal particles at interfaces. *Langmuir* **2009**, *25*, 2718–2728. [CrossRef] [PubMed]

27. Madivala, B.; Vandebril, S.; Fransaer, J.; Vermant, J. Exploiting particle shape in solid stabilized emulsions. *Soft Matter* **2009**, *5*, 1717–1727. [CrossRef]

28. Lewandowski, E.P.; Searson, P.C.; Stebe, K.J. Orientation of a nanocylinder at a fluid interface. *J. Phys. Chem. B* **2006**, *110*, 4283–4290. [CrossRef] [PubMed]

29. Brugarolas, T.; Park, B.J.; Lee, D. Generation of amphiphilic janus bubbles and their behavior at an air–water interface. *Adv. Funct. Mater.* **2011**, *21*, 3924–3931. [CrossRef]

30. Yunker, P.J.; Still, T.; Lohr, M.A.; Yodh, A.G. Suppression of the coffee-ring effect by shape-dependent capillary interactions. *Nature* **2011**, *476*, 308–311. [CrossRef] [PubMed]

31. Vermant, J. Fluid mechanics: When shape matters. *Nature* **2011**, *476*, 286–287. [CrossRef] [PubMed]

32. Ondarçuhu, T.; Fabre, P.; Raphaël, E.; Veyssié, M. Specific properties of amphiphilic particles at fluid interfaces. *J. Phys. France* **1990**, *51*, 1527–1536. [CrossRef]

33. Park, B.J.; Choi, C.-H.; Kang, S.-M.; Tettey, K.E.; Lee, C.-S.; Lee, D. Geometrically and chemically anisotropic particles at an oil–water interface. *Soft Matter* **2013**, *9*, 3383–3388. [CrossRef]

34. Park, B.J.; Choi, C.-H.; Kang, S.-M.; Tettey, K.E.; Lee, C.-S.; Lee, D. Double hydrophilic janus cylinders at an air–water interface. *Langmuir* **2013**, *29*, 1841–1849. [CrossRef] [PubMed]

35. Kang, S.-M.; Kumar, A.; Choi, C.-H.; Tettey, K.E.; Lee, C.-S.; Lee, D.; Park, B.J. Triblock cylinders at fluid–fluid interfaces. *Langmuir* **2014**, *30*, 13199–13204. [CrossRef] [PubMed]

36. Stamou, D.; Duschl, C.; Johannsmann, D. Long-range attraction between colloidal spheres at the air–water interface: The consequence of an irregular meniscus. *Phys. Rev. E* **2000**, *62*, 5263–5272. [CrossRef]

37. Kralchevsky, P.A.; Nagayama, K. Capillary forces between colloidal particles. *Langmuir* **1994**, *10*, 23–36. [CrossRef]

38. Rezvantalab, H.; Shojaei-Zadeh, S. Role of geometry and amphiphilicity on capillary-induced interactions between anisotropic Janus particles. *Langmuir* **2013**, *29*, 14962–14970. [CrossRef] [PubMed]

39. Xie, Q.; Davies, G.B.; Günther, F.; Harting, J. Tunable dipolar capillary deformations for magnetic Janus particles at fluid–fluid interfaces. *Soft Matter* **2015**, *11*, 3581–3588. [CrossRef] [PubMed]

40. Choi, C.-H.; Lee, J.; Yoon, K.; Tripathi, A.; Stone, H.A.; Weitz, D.A.; Lee, C.-S. Surface-tension-induced synthesis of complex particles using confined polymeric fluids. *Angew. Chem. Int. Edit.* **2010**, *49*, 7748–7752. [CrossRef] [PubMed]

41. Jiang, S.; Granick, S. Janus balance of amphiphilic colloidal particles. *J. Chem. Phys.* **2007**, *127*, 161102–161104. [CrossRef] [PubMed]

materials

MDPI

Article

Preparation and Application of Water-in-Oil Emulsions Stabilized by Modified Graphene Oxide

Xiaoma Fei, Lei Xia, Mingqing Chen *, Wei Wei, Jing Luo * and Xiaoya Liu

The Key Laboratory of Food Colloids and Biotechnology, Ministry of Education,
School of Chemical and Material Engineering, Jiangnan University, Wuxi 214122, China;
feixiaoma24@163.com (X.F.); sunnyskyxl@163.com (L.X.); wwei1985@jiangnan.edu.cn (W.W.);
lxy@jiangnan.edu (X.L.)
* Correspondence: mq-chen@jiangnan.edu.cn (M.C.); jingluo19801007@126.com (J.L.);
  Tel./Fax: +86-510-8591-7763 (M.C.)

Academic Editor: To Ngai
Received: 5 June 2016; Accepted: 22 August 2016; Published: 26 August 2016

Abstract: A series of alkyl chain modified graphene oxides (AmGO) with different alkyl chain length and content was fabricated using a reducing reaction between graphene oxide (GO) and alkyl amine. Then AmGO was used as a graphene-based particle emulsifier to stabilize Pickering emulsion. Compared with the emulsion stabilized by GO, which was oil-in-water type, all the emulsions stabilized by AmGO were water-in-oil type. The effects of alkyl chain length and alkyl chain content on the emulsion properties of AmGO were investigated. The emulsions stabilized by AmGO showed good stability within a wide range of pH (from pH = 1 to pH = 13) and salt concentrations (from 0.1 to 1000 mM). In addition, the application of water-in-oil emulsions stabilized by AmGO was investigated. AmGO/polyaniline nanocomposite (AmGO/PANi) was prepared through an emulsion approach, and its supercapacitor performance was investigated. This research broadens the application of AmGO as a water-in-oil type emulsion stabilizer and in preparing graphene-based functional materials.

Keywords: modified graphene oxide; Pickering emulsion; water-in-oil emulsions; nanocomposite

1. Introduction

Owing to its water dispersibility, graphene oxide (GO), as a derivative of graphene, has many applications such as transparent conductors, sensors, polymer composites and energy storage [1–7]. GO is typically synthesized by chemical oxidation and exfoliation of graphite powders using strong oxidizing agents. After oxidation, the GO sheets are derivatized by a 2D lattice of partially broken sp^2-bonded carbon networks with phenol, hydroxyl, and epoxide groups on the basal planes and carboxylic acid groups at the edges. Such a structure makes GO an amphiphile with a large hydrophobic basal plane and hydrophilic edges [8–11].

Since the pioneering works of Huang devoted to investigated the interfacial activity of GO at air-water, liquid-liquid, and liquid-solid interfaces and proved GO can act as a colloidal surfactant, many interesting works in the last several years have been focused on making GO at the water/oil interface [12–19]. He et al. investigated extensively the effects of different conditions such as type of oil, the sonication time, the GO concentration, the oil/water ratio, and the pH value on the properties of the Pickering emulsions stabilized by GO [16]. Creighton et al., presented the thermodynamic analysis of the behavior of 2D materials such as GO at liquid—liquid interfaces with applications in emulsification [17]. McCoy et al. demonstrated that oil/water emulsions stabilized by GO could be flocculated by either an increase or a decrease in pH. Such a flocculation was fully reversible at highly acidic pH and irreversible at high pH [18]. More than adsorption to the oil/water interface,

Sun and co-workers reported that the surfactancy of GO can be enhanced by a block copolymer as a ligand. They added the copolymer ligand into oil, so GO can be trapped at water/oil interface and jammed into a solid thin film. Its kinetics were studied, but the emulsion properties of GO were not investigated and application of this GO thin film was not very extensive [19].

Making GO at the interface could provide guidance for the fabrication of graphene-based functional materials with specific structure and performance. Because the Pickering emulsions stabilized by GO can be used as a soft template to design new functional hybrid materials [20–26]. It should be pointed out that the graphene-based composites were usually synthesized through an "emulsion polymerization" or "suspension polymerization" since the graphene moieties very could well stabilize the monomer. Xie et al. reported the preparation of polystyrene (PS) particles via Pickering emulsion polymerization using GO as the stabilizer [27]. A similar "Pickering emulsion polymerization" approach was also reported by Kattimuttathu and co-workers; they used a so-called "seeded" emulsion polymerization to developed GO/PS nanocomposites [28]. Dao and co-workers prepared a microsphere of poly (methyl methacrylate) (PMMA)/graphene composite with a core–shell structure by Pickering suspension polymerization. First, they used GO to stabilize MMA (methylmethacrylate) in water, and then activated the incorporated initiator, AIBN, to turn the emulsified droplets into solid PMMA beads [29].

As stated above, Pickering emulsion stabilized by GO tends to be oil-in-water (O/W) type, which is not suitable for applications where the water-in-oil (W/O) type emulsion is needed. Therefore, some research has been focusing on the preparation of graphene-based W/O emulsion [25,30–33]. Woltornist and co-workers demonstrated the use of graphene/graphite to stabilize water in styrene/divinylbenzene and then curing the continuous monomer phase. After evaporation of the water, a semi-open cell carbon (graphene) network with high electrical conductivities was produced [32]. Zheng and co-workers reported when GO was modified by the cationic surfactant cetyltrimethylammonium bromide (CTAB), it could stabilize a W/O Pickering emulsion. They used the Pickering high internal phase emulsion stabilized by GO as templates and successfully prepared macroporous polymer/GO composites with a high specific surface area of about 490 $m^2 \cdot g^{-1}$ [25].

In this paper, a series of alkyl chain modified GO (AmGO) was synthesized, and then they were served as efficient W/O type emulsion stabilizers. The effects of alkyl chain length and alkyl chain content on the emulsion properties of AmGO were studied. Then the application of W/O emulsion stabilized by AmGO was investigated. The emulsion stabilized by AmGO was used as a soft template to prepared AmGO/polyaniline nanocomposite (AmGO/PANi). The AmGO/PANi exhibited a different nanostructure and a better supercapacitor performance than the common graphene oxide/polyaniline nanocomposite.

2. Experimental Materials

Natural flake graphite was obtained from Qingdao Zhongtian Co., Ltd. (Qingdao, China) with a particle size of 400 meshes. All other chemicals are of analytical grade and used as received without further purification. Deionized (DI) water was used for the washings.

2.1. Preparation of Graphene Oxide (GO)

GO was prepared with the Hummers method and purified. Graphite (5 g) and NaNO$_3$ (2.5 g) were mixed with 120 mL of H$_2$SO$_4$ (95%) in a 500 mL flask. The mixture was stirred for 30 min within an ice bath. While maintaining vigorous stirring, potassium permanganate (15 g) was added to the suspension. The rate of addition was carefully controlled to keep the reaction temperature lower than 20 °C. The ice bath was then removed, and the mixture was stirred at room temperature overnight. As the reaction progressed, the mixture gradually became pasty, and the color turned into light brownish. At the end, 150 mL of H$_2$O was slowly added to the pasty with vigorous agitation. The reaction temperature was rapidly increased to 98 °C with effervescence, and the color changed to yellow. The diluted suspension was stirred at 98 °C for one day. Then, 50 mL of 30% H$_2$O$_2$ was added to the mixture. For purification, the mixture was washed by rinsing and centrifugation with 5% HCl and water for several times. Then, GO was dried under vacuum and obtained as a gray powder.

2.2. Preparation of Alkyl Chain Modified Graphene Oxide (AmGO)

GO (0.3 g) was dispersed and exfoliated in 150 mL deionized water via ultrasonication. The resulting suspension was mixed with the solution with a certain amount of alkyl amine (n-hexylamine, dodecylamine or octadecylamine) in 50 mL ethanol in a three-neck flask. The mixture was refluxed with mechanical stirring for 20 h at 90 °C. The precipitate was filtrated with a PP membrane with an average pore size of 220 nm. The filtrated powder was redispersed in 50 mL ethanol via ultrasonication for 2 min and then refiltrated. The rinsing-filtration process was repeated for 4 times to remove the physically absorbed alkyl amine. Finally, the mixture was dried at 60 °C under vacuum.

2.3. Preparation of Pickering Emulsions Stabilized by Amgo

AmGO was dispersed in toluene via ultrasonication. The resulting dispersion was mixed with a certain amount of deionized water at different pH values or salt concentrations in glass vessels at room temperature. Then the mixtures were homogenized at 3000 rpm for 1 min by a XHF-DH-speed dispersator homogenizer (1 cm head) (Scientz, Ningbo, China). The type of emulsions was determined by drop test.

2.4. Preparation of Graphene/Polyaniline Nanocomposite

For the synthesis of AmGO/PANi, AmGO and aniline were dispersed in toluene via ultrasonication in glass vessels at room temperature. 10 mL of such dispersion was mixed with 1 mL of 1 M HCl contain a certain amount of APS. The mixture was homogenized at 3000 rpm for 1 min and then reacted at 0 °C. After reaction, the product was vacuum filtered and washed with acetone and deionized water to remove excess acid, possible oligomers. The obtained product was dried completely at 50 °C. For the synthesis of GO/PANi, aniline and GO were dispersed in water via ultrasonication. Then contain a certain amount of HCl solution and APS were added into the mixture. The mixture was reacted at 0 °C. After reaction, the product was vacuum filtered and washed with acetone and deionized water.

2.5. Characterization

The photos of the emulsions stabilized by alkyl chain modified GO were recorded with a digital camera (COOLPIX P100, Nikon, Tokyo, Japan). The optical micrographs of the prepared emulsions on transparent glass slides were taken using an optical microscope (DM-BA450, Motic China Group Co., Ltd., Xiamen, China). The average sizes and size distributions of the emulsion droplets were determined based on the images of the emulsion droplets using Nano Measurer 1.2 software. The volume fraction of the stable emulsion was calculated by dividing the height of the emulsion phase by the total height of the initial emulsion. Water contact angle measurements were performed at ambient conditions using an OCA15EC goniometer with a charge-coupled device camera equipped for image capture (Dataphysics, Beijing, China). The GO and AmGO powder samples were first made into a film by a tableting machine. Typically, 0.5 g sample was used and the pressure was kept at 15 MPa for 2 min. The obtained film was around 0.1 mm thick. Static contact angle measurements were performed by placing a 2 μL deionized water droplet on the surface of the prepared film. The axisymmetric-drop-shape analysis profile (ADSA-P) method was used for estimating the contact angle of water on the film surface. The SEM image was taken in an S-4800 field emission scanning electron microscope (Hitachi, Tokyo, Japan). Raman spectra were recorded using a Renishaw in Via Raman Microscope (Renishaw, Gloucestershire, UK) operating at 785 nm with a charge-coupled device detector. FT-IR spectroscopy were collected on a Nicolet IS-50 FT-IR spectrometer equipped with a Smart OMNI sampler with a high purity Ge crystal and diamond crystal (Nicolet, Green Bay, WI, USA). Electrochemical characterization was carried out in a conventional three-electrode cell system with a nanocomposite modified glassy carbon electrode (GCE) was used as the working electrode. First, GCE was carefully polished with alumina powders (0.3 and 0.05 mm) on a polishing cloth, then sonicated in ethanol and dried at room temperature. To prepare working electrode, typically, 50 mg of nanocomposite was homogeneously dispersed in 50 mL of water under ultrasonication to obtain the dispersion with concentration of 1.0 mg·mL^{-1}. Then 10 µL of the obtained dispersion was dropped onto the GCE surface, and then dried at room temperature to form the nanocomposite modified GCE. Electrochemical measurement was performed on a CHI660e electrochemical workstation (Shanghai Chenghua Instrument Company, Shanghai, China) in a conventional three-electrode cell system. The nanocomposite modified GCE, a saturated calomel electrode (SCE) and a Pt wire were used as working electrode, reference electrode and counter electrode, respectively. Galvanostatic charge/discharge curve was measured in a voltage from 0 to 0.8 V with 1.0 M H$_2$SO$_4$ as the electrolyte.

3. Results and Discussion

3.1. Basic Characterization of AmGO

The GO used in this work was prepared from purified natural graphite via a modified Hummers method [34]. The AmGO was synthesized using different alkyl amine to reduce GO according to previous report [35,36]. Here, three alkyl amine with different alkyl chain length (n-hexylamine (6-C), dodecylamine (12-C) or octadecylamine (18-C)) were utilized to modified GO. AmGO with different GO and octadecylamine ratios were also synthesised. For easy expression, the alkyl chain length in AmGO was labeled as X, and the weight ratio between GO and alkyl amine in AmGO was labeled as Y. Thus, samples with different alkyl chain length and alkyl chain content were labeled as AmGOX-Y. For example, AmGO18-1 represented for the octadecylamine (18-C) modified GO with the weight ratio of 1:1.

The basic characterization is given in Figure 1. Figure 1a shows the FTIR spectra of GO and AmGO18-1. After modified with alkyl amine, the bands of GO at 1725 cm^{-1} and 1581 cm^{-1} which should be assigned to the C=O carboxyl stretching vibration band and C=C in aromatic ring were disappeared. A new band at 1541–1603 cm^{-1} assigned to the carboxylic acid salt (COO^{-}) asymmetric stretch mode was observed. Peaks at 2918, 2848 and 720 cm^{-1} are due to the –CH$_2$– in alkyl chain in the spectrum of AmGO18-1. The result of Raman was given in Figure 1b. Both the two samples showed two bands in Raman spectra, D band and G band, which suggested that the skeleton structure of GO remained in AmGO18-1 after modification. However, after modification, the I$_D$/I$_G$ value decreased, which means GO was reduced in this process. This result is in consistent with previous literatures, which were also modifying the GO with alkyl amine and proved GO could be reduced [32,35]. Both the FTIR and Raman results demonstrate the successful preparation of AmGO.

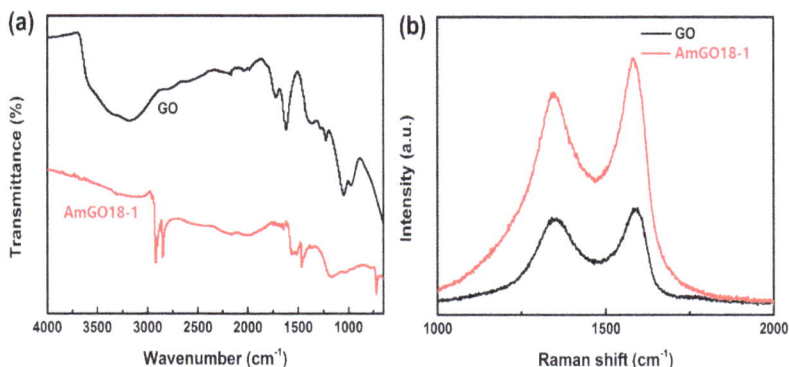

Figure 1. (**a**) FTIR spectra of GO and AmGO18-1; (**b**) Raman spectra of GO and AmGO18-1.

3.2. Effect of Alkyl Chain Length on the Emulsion Properties of AmGO

As is known, GO is highly hydrophilic owning to the oxygen-containing functional groups, such as phenol, hydroxyl, epoxide and especially the edge carboxylic acid groups. So GO tends to stabilize O/W type emulsion. In most cases, the emulsion properties of a stabilizer is largely related to its amphipathicity. Here, different alkyl amines were utilized to introduce hydrophobic alkyl chains on GO surface to adjust its amphipathicity. As all the AmGO could disperse in toluene, toluene was chosen as oil phase to investigate the emulsion properties. First, the effect of alkyl chain length on the emulsion properties was investigated. The results were given in Figure 2. Clearly, all the three AmGO samples (AmGO6-1, AmGO12-1, AmGO18-1) were observed to efficiently stabilize emulsions from toluene/water mixtures. The emulsions stayed stable for at least several months but sedimented, leaving a clear oil phase above the emulsion. Such a result showed the good stability of emulsion stabilized by AmGO. Compared with the emulsion stabilized by GO, which is O/W type, all the emulsions stabilized by AmGO here belonged to W/O type. This could be explained by the following: it is well known that if a stabilizer is more soluble in the oil phase, it is easily to obtain a W/O emulsion. In this study, GO is more soluble in water, so it can emulsify toluene in water. For AmGO, after being modified with alkyl chain, the AmGO becomes more hydrophobic and could easily disperse in the toluene, which is the oil phase. Thus, AmGO tends to stabilize the W/O emulsion. The water contact angles were measured to illustrate such an increase in hydrophobicity (Figure 3). The water contact angle of GO was 50°, which is much smaller than 90°, showing that GO is quite hydrophilic and tend to obtain O/W type emulsion. For AmGO, the water contact angles were much higher than that of pure GO and increased with raising the length of alkyl chain. What should be pointed out is that the water contact angle of AmGO6-1 was 81°, which is still less than 90°. As the length of the alkyl chain increases, both AmGO12-1 and AmGO18-1 showed water contact angles higher than 90° and possessed enough hydrophobicity, leading to good emulsion properties. In addition, all the emulsions stabilized by AmGO reached a stable state very quickly and the volume fractions of the residual emulsion were more than 60% (Figure 2d).

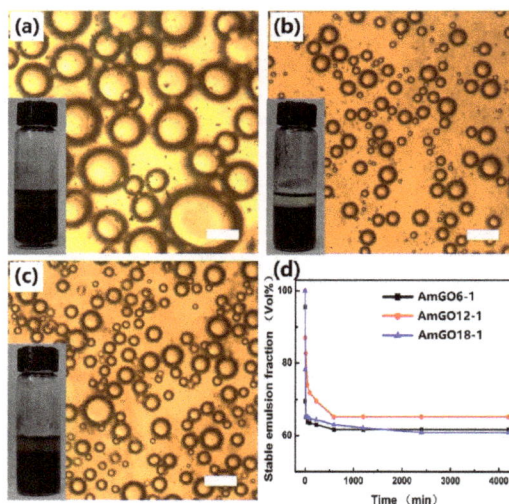

Figure 2. Optical micrographs and photographs 72 h after preparation of water-in-toluene Pickering emulsions stabilized by GO modified with different organic amine: (**a**) AmGO6-1; (**b**) AmGO12-1 and (**c**) AmGO18-1; (**d**) Change in the volume fraction of the residual emulsion as a function of time for the emulsions stabilized by GO modified with different organic amine. Concentration: 1 mg/mL. Toluene/water ratio: 1:1.

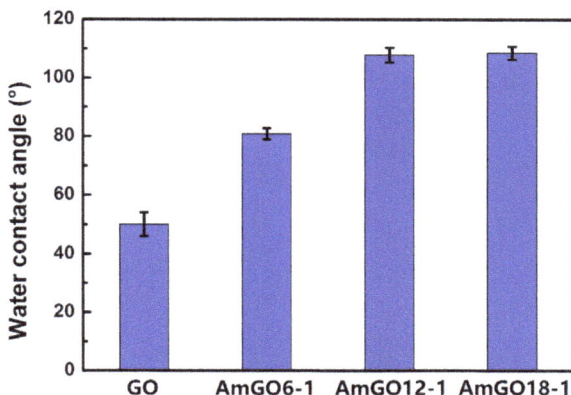

Figure 3. Water contact angles of GO, AmGO6-1, AmGO12-1 and AmGO18-1.

3.3. Effect of the Grafting Content of Alkyl Chain on the Emulsion Properties of AmGO

The amphipathicity of AmGO could also be tuned by the grafting content of alkyl chain. Here, four AmGO with different GO and octadecylamine ratios were synthesized to illustrate the effect of the grafting content on emulsion properties of AmGO. As expected, the hydrophobicity increased with increasing grafting content of octadecylamine. This could be witnessed by the results of water contact angles (Figure 4a). The water contact angles of all the AmGO were larger than 90° and increased with the content of octadecylamine. The emulsion properties of these four AmGO were given in Figure 5. All the four AmGO could form water/toluene type emulsions. Obviously, droplets of the emulsion stabilized by AmGO18-1 were very uniform and the stable emulsion fraction reached 65%, which was much higher than those of other three AmGO (Figure 4b). For the emulsion stabilized by AmGO18-0.5, the morphology of the emulsion droplet was irregular, and free AmGO18-0.5 sheets were observed. This could be explained by the fact that the content of octadecylamine on GO is too low, so AmGO18-0.5 could not disperse in toluene efficiently and settle down in the bottom, leading to poor emulsion properties. For the emulsions stabilized by AmGO18-2 and AmGO18-4, their emulsion droplet sizes were similar to that of AmGO18-1. However, the stable emulsion fractions were about 25%, which was much lower than that of AmGO18-1. The reason for this observation is that there is too many hydrophobic octadecylamine chains on GO surface for AmGO18-2 and AmGO18-4, which makes the materials too hydrophobic and thus decreases their active surface.

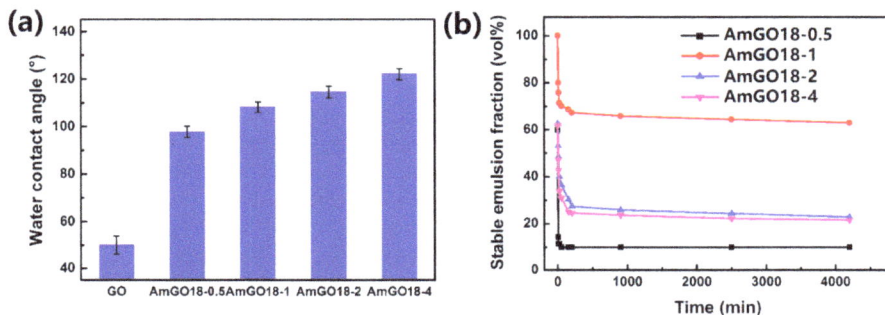

Figure 4. (**a**) Water contact angles of GO and AmGO; (**b**) Change in the volume fraction of the residual emulsion as a function of time for the emulsions stabilized by AmGO18-0.5, AmGO18-1, AmGO18-2 and AmGO18-4. AmGO concentration: 1 mg/mL. Toluene/water ratio: 1:1.

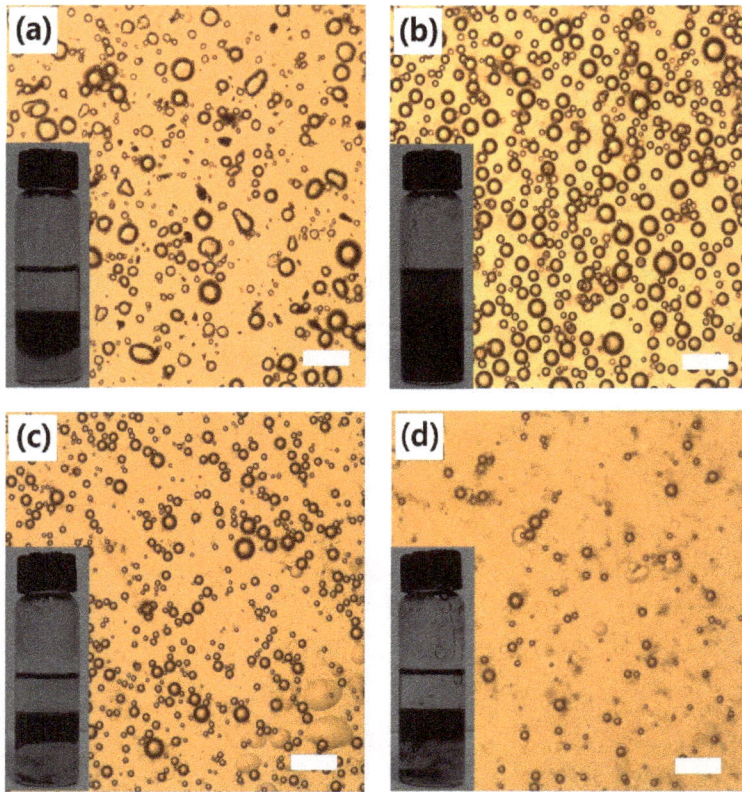

Figure 5. Optical micrographs and photographs after 72 h preparation of Pickering emulsions stabilized by AmGO: (**a**) AmGO18-0.5; (**b**) AmGO18-1; (**c**) AmGO18-2 and (**d**) AmGO18-4. AmGO concentration: 1 mg/mL. Toluene/water ratio: 1:1. Scale bar: 100 μm.

3.4. Effect of pH Value and NaCl Concentration on the Emulsion Properties of AmGO

The amphiphilicity of GO, as pointed out above, largely depends on the edge carboxyl groups. Thus, the emulsion properties of GO is strongly influenced by varying the pH values due to protonation and deprotonation of the edge carboxyl groups. McCoy reported the pH-dependent dispersion and flocculation of the emulsion stabilized by GO through either an increase or a decrease in pH [18]. Here the effect of pH value on the emulsion properties of AmGO18-1 was investigated, and the results are shown in Figure 6 and Figure S1. All the emulsions stabilized by AmGO18-1 at different pH values showed similar emulsion droplet size and similar stable emulsion volume. Such a result means that the emulsion properties of AmGO18-1 was not influenced by pH, which is in contrast with GO. This is because GO was partly reduced, so the oxygen-containing functional groups on GO decreased, which significantly decreased the sensitivity to pH value. However, it should be noted that the emulsion droplet size become smaller and more uniform and the toluene phase became transparent in basic conditions. This means that nearly all the AmGO18-1 in the dispersion was associated with the emulsification at high pH.

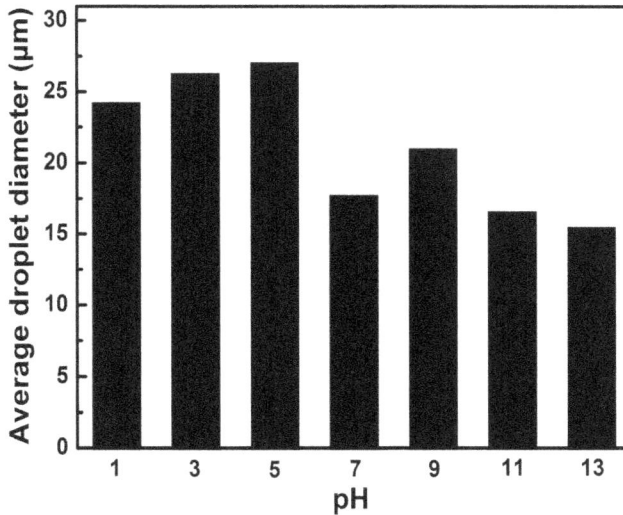

Figure 6. Average droplet diameter of Pickering emulsions stabilized by AmGO18-1 as a function of pH values. AmGO18-1 concentration: 1 mg/mL. Toluene/water ratio: 1:1. Scale bar: 100 μm.

It was reported that emulsion stabilized by GO was also affected by electrolytes, such as NaCl, which affect the aggregation behavior of the GO dispersion. This is because both the carboxylic acid and hydroxyl groups on GO surface are candidates for charge screening when the salt is added. For the emulsions stabilized by AmGO18-1, as expected, the effect of NaCl concentration on the emulsion droplet size and on the stable emulsion volume was not distinct. All the emulsions showed similar stable emulsion volumes and no flocculation was observed (Figure S2).

3.5. Effect of AmGO Concentration on the Emulsion Properties of AmGO

Particle concentration is another important factor in the formation of Pickering emulsion, which has an influence on the emulsion stability and on the average droplet size. The microscopy images and distribution of droplet diameters for emulsions prepared with different AmGO12-1 concentrations were shown in Figure 7 and Figure S3. For AmGO18-1 concentration was 0.1 mg·mL^{-1}, there was almost no emulsion formed. When the concentration of AmGO18-1 reached 0.2 mg·mL^{-1}, a uniform W/O type emulsion was formed and there was no separated water phase. This obvious change could be explained as following. AmGO has a very large specific surface area. The increase of AmGO18-1 concentration within a certain range significantly results in an increase of the total surface area. Such an increase of the surface area of AmGO18-1 causes a reduction in the free energy and makes the system more stable, thus producing good emulsion properties. For concentrations greater than 0.2 mg·mL^{-1}, as expected, the average droplet size decreased and the volume fraction of the residual emulsion increased with increasing AmGO18-1 concentration.

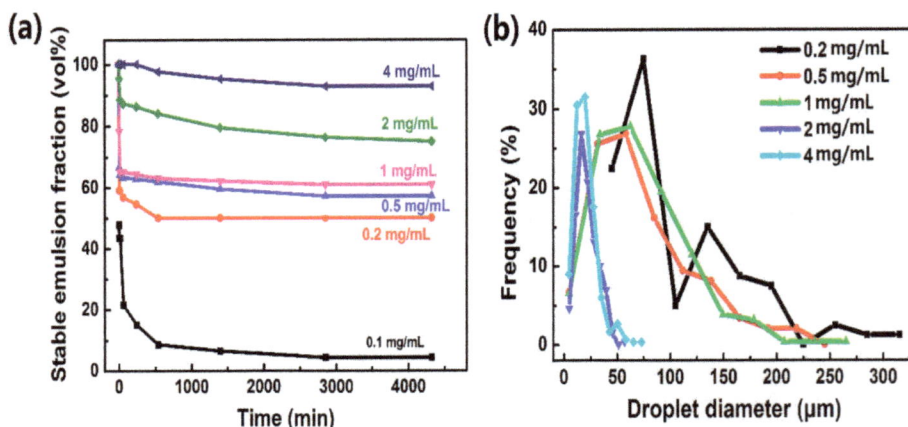

Figure 7. (**a**) Change in the volume fraction of the residual emulsion as a function of time for the emulsions stabilized by different concentrations of AmGO18-1; (**b**) Distribution of droplet diameters after the preparation of water-in-toluene emulsions at different AmGO18-1 concentrations.

3.6. Preparation of AmGO/PANi and Its Supercapacitor Performance

The main application of GO used as an emulsion stabilizer is to design novel functional graphene-based materials with specific structure and performance using emulsion stabilized by GO as a soft template. Much research has focused on preparing GO/PS material using emulsion polymerization [12,34,35]. However, as was mentioned above, emulsion stabilized by GO was highly influenced by pH value, which limited the choice of monomer, for example aniline. As we all know, polyaniline (PANi) is a typical conducting polymer which has been used in various fields [36–38]. Its synthesis is typically performed using oxidation polymerization in an acid environment. We had tried to prepare GO/PANi nanocomposite through emulsion polymerization using O/W emulsion stabilized by GO as a soft template (GO solution as water phase and toluene dissolved with aniline monomer as oil phase). The system could form O/W emulsion. However, after adjusting the pH value and adding APS to initiate polymerization, the reaction system flocculated rapidly, leaving brown precipitate in the bottle. This is because emulsion stabilized by GO is unstable under low pH values and became flocculate. As is demonstrated, emulsion stabilized by AmGO is very stable within a wide range of pH. Thus, AmGO/PANi nanocomposite was also prepared through emulsion polymerization using the AmGO18-1 as emulsion stabilizer (HCl solution dissolved with APS as water phase and toluene dispersed with AmGO18-1 and aniline monomer as oil phase). As expected, the reaction system remained stable during the whole polymerization progress. To understand the difference in structure between the graphene/polyaniline composite synthesized through the emulsion strategy and the common solution method, we also prepared GO/PANi nanocomposite using a common solution polymerization. It is easy to realize that for the GO/PANi, the polymerization of aniline followed a surface-initiated polymerization mechanism. Unlike GO/PANi, the obtained AmGO/PANi nanocomposite synthesised through an emulsion strategy showed an interfacial polymerization mechanism. Such a difference could be demonstrated by the results of TEM. It can be seen that the PANi nanofibers and nano-spheres were formed and randomly attached on the AmGO surface. The diameter of PANi nanofibers was around 39 nm [39,40]. However, the morphology of GO/PANi was very different from that of AmGO/PANi, i.e., the GO sheet were homogeneously covered by PANi layer and no specific nanostructure was observed (Figure 8b). This revealed a surface-initiated polymerization process.

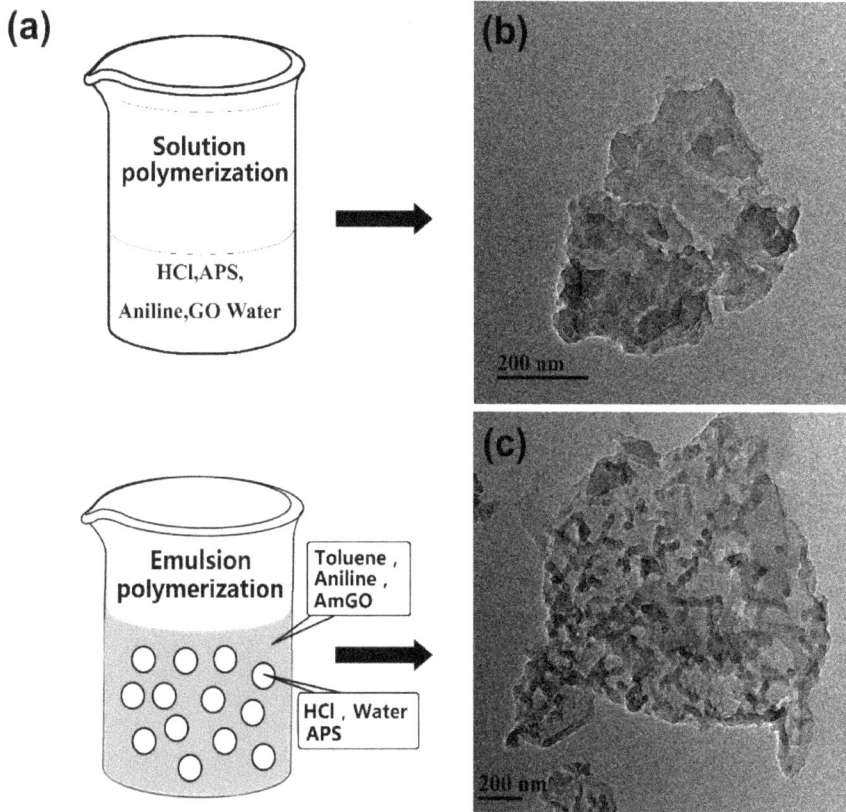

Figure 8. (a) Illustration of the solution polymerization and emulsion polymerization process; TEM images of (b) GO/PANi and (c) AmGO/PANi.

The capacitive behavior difference between AmGO/PANi and GO/PANi was confirmed by galvanostatic charge/discharge curves and the curves at a current density of 1 A·g^{-1} are shown in Figure 9a. Apparently, the AmGO/PANi prepared from emulsion polymerization method exhibited a better supercapacitor performance than GO/PANi. The specific capacitance of AmGO/PANi reached 585 F·g^{-1} at a current density of 1 A·g^{-1}, which was much higher than that of GO/PANi (405 F·g^{-1}). The electrochemical stability of AmGO/PANi and GO/PANi nanocomposites were also evaluated by consecutive charge/discharge cycles at a current density of 2 A·g^{-1}. As shown in Figure 9b, the AmGO/PANi exhibited a better cycling stability than GO/PANi. About 67.8% of the initial capacitance could be maintained after 1000 cycles at a current density of 2 A·g^{-1}, much higher than that of GO/PANI (49%). Such a better specific capacitance and cycling life may be due to the special structure in the nanocomposite obtained through emulsion polymerization method using the AmGO as W/O emulsifier.

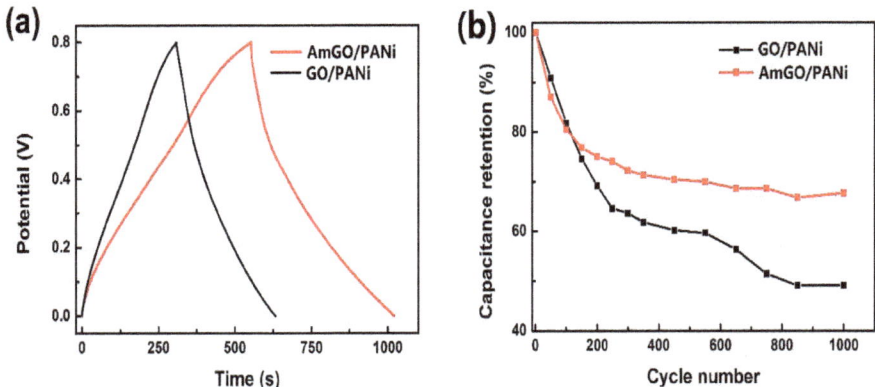

Figure 9. (**a**) charge-discharge curves of AmGO/PANi and GO/PANi at a current density of 1 A·g^{-1} and (**b**) cycling stability of AmGO/PANi and GO/PANi at a current density of 2 A·g^{-1}.

4. Conclusions

In conclusion, we have shown the preparation and application of W/O type emulsions stabilized by AmGO. The effects of alkyl chain length and alkyl chain content on the emulsion properties of AmGO were investigated. The results showed that the emulsion properties of AmGO significantly depend on its amphipathicity. AmGO18-1 exhibited the best emulsion properties with a good emulsion stability within a wide range of pH (from pH = 1 to pH = 13) and salt concentrations (from 0.1 to 1000 mM). AmGO/PANi was synthesized using an emulsion approach. Compared with GO/PANi, AmGO/PANi exhibited a different nanostructure and a better supercapacitor performance. The specific capacitance of AmGO/PANi reached 585 F·g^{-1} at a current density of 1 A·g^{-1}. This research broadens the application of AmGO as a W/O type emulsion stabilizer and in preparing graphene-based functional materials.

Supplementary Materials: The following are available online at www.mdpi.com/1996-1944/9/9/731/s1. Figure S1: Optical micrographs and photographs after 72 h preparation of Pickering emulsions stabilized by AmGO18-1 at different pH values: (a) pH = 1; (b) pH = 3; (c) pH =5; (d) pH = 7; (e) pH = 9; (f) pH = 11 and (g) pH = 13. AmGO18-1 concentration: 1 mg/mL. Toluene/water ratio: 1:1. Scale bar: 100 μm; Figure S2: Photographs 72 h after preparation of Pickering emulsions stabilized by AmGO18-1 with different NaCl concentrations. The concentrations of NaCl (mM) are (a) 0.1; (b) 1; (c) 10; (d) 20; (e) 50; (f) 100; (g) 300; (h) 500; and (i) 1000. AmGO18-1 concentration: 1 mg/mL. Toluene/water ratio: 1:1; Figure S3: Optical micrographs and photographs after 72 h preparation of Pickering emulsions stabilized by AmGO18-1 at different AmGO18-1 concentrations: (a) 0.1; (b) 0.2; (c) 0.5; (d) 1; (e) 2; (f) 4 mg/mL. Toluene/water ratio: 1:1. Scale bar: 100 μm.

Acknowledgments: We are grateful for the financial support from the National Natural Science Foundation of China (No. 51103064 and 51173072), the Fundamental Research Funds for the Central Universities (JUSRP 51305A), MOE & SAFEA for the 111 Project (B13025).

Author Contributions: X.F. and J.L. conceived and designed the experiments; X.F. and L.X. performed the experiments; X.F. and W.W. analyzed the data; X.F., M.C. and X.L. contributed reagents/materials/analysis tools; X.F. wrote the paper. All authors contributed to the technical review of the manuscript.

Conflicts of Interest: The authors declare no conflict of interest.

References

1. Wassei, J.K.; Kaner, R.B. Graphene, a promising transparent conductor. *Mater. Today* **2010**, *13*, 52–59. [CrossRef]

2. Jiang, H.J. Chemical preparation of graphene-based nanomaterials and their applications in chemical and biological sensors. *Small* **2011**, *7*, 2413–2427. [CrossRef] [PubMed]

3. Potts, J.R.; Dreyer, D.R.; Bielawski, C.W.; Ruoff, R.S. Graphene-based polymer nanocomposites. *Polymer* **2011**, *52*, 5–25. [CrossRef]

4. Huang, X.; Qi, X.Y.; Boey, F.; Zhang, H. Graphene-based composites. *Chem. Soc. Rev.* **2012**, *41*, 666–686. [CrossRef] [PubMed]
5. Stoller, M.D.; Park, S.; Zhu, Y.; An, J.; Ruoff, R.S. Graphene-based ultracapacitors. *Nano Lett.* **2008**, *8*, 3498–3502. [CrossRef] [PubMed]
6. Fowler, J.D.; Allen, M.J.; Tung, V.C.; Yang, Y.; Kaner, R.B.; Weiller, B.H. Practical chemical sensors from chemically derived graphene. *ACS Nano* **2009**, *3*, 301–306. [CrossRef] [PubMed]
7. Xu, B.; Yue, S.F.; Sui, Z.Y.; Zhang, X.T.; Hou, S.S.; Cao, G.P.; Yang, Y.S. What is the choice for supercapacitors: Graphene or graphene oxide? *Energy Environ. Sci.* **2011**, *4*, 2826–2830. [CrossRef]
8. Li, D.; Muller, M.B.; Gilje, S.; Kaner, R.B.; Wallace, G.G. Processable aqueous dispersions of graphene nanosheets. *Nat. Nanotechnol.* **2008**, *3*, 101–105. [CrossRef] [PubMed]
9. Li, D.; Kaner, R.B. Graphene-Based Materials. *Science* **2008**, *320*, 1170–1171. [CrossRef] [PubMed]
10. Park, S.; Ruoff, R.S. Chemical methods for the production of graphenes. *Nat. Nanotechnol.* **2009**, *4*, 217–224. [CrossRef] [PubMed]
11. Erickson, K.; Erni, R.; Lee, Z.; Alem, N.; Gannett, W.; Zettl, A. Determination of the local chemical structure of graphene oxide and reduced graphene oxide. *Adv. Mater.* **2010**, *40*, 4467–4472. [CrossRef] [PubMed]
12. Texter, J. Graphene oxide and graphene flakes as stabilizers and dispersing aids. *Curr. Opin. Colloid Interface Sci.* **2015**, *20*, 454–464. [CrossRef]
13. Luo, J.Y.; Cote, L.J.; Tung, V.C.; Tan, A.T.L.; Goins, P.E.; Wu, J.S.; Huang, J.X. Graphene oxide nanocolloids. *J. Am. Chem. Soc.* **2010**, *132*, 17667–17669. [CrossRef] [PubMed]
14. Cote, L.J.; Kim, J.Y.; Tung, V.C.; Luo, J.Y.; Kim, F.; Huang, J.X. Graphene oxide as surfactant sheets. *Pure Appl. Chem.* **2011**, *83*, 95–110. [CrossRef]
15. Kim, J.Y.; Cote, L.J.; Kim, F.; Yuan, W.; Shull, K.R.; Huang, J.X. Graphene oxide sheets at interfaces. *J. Am. Chem. Soc.* **2010**, *132*, 8180–8186. [CrossRef] [PubMed]
16. He, Y.Q.; Wu, F.; Sun, X.Y.; Li, R.Q.; Guo, Y.Q.; Li, C.B.; Zhang, L.; Xing, F.B.; Wang, W.; Gao, J.P. Factors that affect pickering emulsions stabilized by graphene oxide. *ACS Appl. Mater. Interfaces* **2013**, *5*, 4843–4855. [CrossRef] [PubMed]
17. Creighton, M.A.; Ohata, Y.; Miyawaki, J.; Bose, A.; Hurt, R.H. Two-dimensional materials as emulsion stabilizers: Interfacial thermodynamics and molecular barrier properties. *Langmuir* **2014**, *30*, 3687–3696. [CrossRef] [PubMed]
18. McCoy, T.M.; Pottage, M.J.; Tabor, R.F. Graphene oxide-stabilized oil-in-water emulsions: pH-controlled dispersion and flocculation. *J. Phys. Chem. C* **2014**, *118*, 4529–4535. [CrossRef]
19. Sun, Z.W.; Feng, T.; Russell, T.P. Assembly of graphene oxide at water/oil interfaces: Tessellated nanotiles. *Langmuir* **2013**, *29*, 13407–13413. [CrossRef] [PubMed]
20. Tang, M.Y.; Wang, X.R.; Wu, F.; Liu, Y.; Zhang, S.; Pang, X.B.; Li, X.X.; Qiu, H.X. Au nanoparticle/graphene oxide hybrids as stabilizers for Pickering emulsions and Au nanoparticle/graphene oxide@ polystyrene microspheres. *Carbon* **2014**, *71*, 238–248. [CrossRef]
21. Gudarzi, M.M.; Sharif, F. Self assembly of graphene oxide at the liquid–liquid interface: A new route to the fabrication of graphene based composites. *Soft Matter* **2011**, *7*, 3432–3440. [CrossRef]
22. Moghaddam, S.Z.; Saburya, S.; Sharif, F. Dispersion of rGO in polymeric matrices by thermodynamically favorable self-assembly of GO at oil–water interfaces. *RSC Adv.* **2014**, *4*, 8711–8719. [CrossRef]
23. Thickett, S.C.; Zetterlund, P.B. Preparation of composite materials by using graphene oxide as a surfactant in Ab initio emulsion polymerization systems. *ACS Macro Lett.* **2013**, *2*, 630–634. [CrossRef]
24. Sun, J.; Bi, H. Pickering emulsion fabrication and enhanced supercapacity of graphene oxide-covered polyaniline nanoparticles. *Mater. Lett.* **2012**, *81*, 48–51. [CrossRef]
25. Zheng, Z.; Zheng, X.H.; Wang, H.T.; Du, Q.G. Macroporous graphene oxide–polymer composite prepared through Pickering high internal phase emulsions. *ACS Appl. Mater. Interfaces* **2013**, *5*, 7974–7982. [CrossRef] [PubMed]
26. Zhang, F.F.; Zhang, X.B.; Dong, Y.H.; Wang, L.M. Enhancing electrical conductivity of rubber composites by constructing interconnected network of self-assembled graphene with latex mixing. *J. Mater. Chem.* **2012**, *22*, 10464–10468. [CrossRef]
27. Xie, P.F.; Ge, X.P.; Fang, B.; Li, Z.; Liang, Y.; Yang, C.Z. Pickering emulsion polymerization of graphene oxide-stabilized styrene. *Colloid Polym. Sci.* **2013**, *7*, 1631–1639. [CrossRef]

28. Kattimuttathu, S.I.; Krishnappan, C.; Vellorathekkaepadil, V.; Nutenki, R.; Mandapati, V.R.; Cernik, M. Synthesis, characterization and optical properties of graphene oxide–polystyrene nanocomposites. *Polym. Adv. Technol.* **2015**, *26*, 214–222.

29. Dao, T.D.; Efdenedelger, G.; Jeong, H.M. Water-dispersible graphene designed as a Pickering stabilizer for the suspension polymerization of poly(methyl methacrylate)/graphene core–shell microsphere exhibiting ultra-low percolation threshold of electrical conductivity. *Polymer* **2014**, *55*, 4709–4719. [CrossRef]

30. Hummers, W.S.; Offeman, R.E. Preparation of graphitic oxide. *J. Am. Chem. Soc.* **1958**, *80*, 1339–1339. [CrossRef]

31. Li, W.; Tang, X.Z.; Zhang, H.B.; Jiang, Z.G.; Yu, Z.Z.; Du, X.S.; Mai, Y.W. Simultaneous surface functionalization and reduction of graphene oxide with octadecylamine for electrically conductive polystyrene composites. *Carbon* **2011**, *49*, 4724–4730. [CrossRef]

32. Woltornist, S.J.; Carrillo, J.M.Y.; Xu, T.O.; Dobrynin, A.V.; Adamson, D.H. Polymer/pristine graphene based composites: From emulsions to strong, electrically conducting foams. *Macromolecules* **2015**, *48*, 687–693. [CrossRef]

33. Mohaghedgh, N.; Tasviri, M.; Rahimi, E.; Gholami, M.R. A novel p–n junction $Ag_3PO_4/BiPO_4$-based stabilized Pickering emulsion for highly efficient photocatalysis. *RSC Adv.* **2015**, *5*, 12944–12955. [CrossRef]

34. Patolea, A.S.; Patoleb, S.P.; Kanga, H.; Yoob, J.B.; Kima, T.H.; Ahn, J.H. A facile approach to the fabrication of graphene/polystyrene nanocomposite by in situ microemulsion polymerization. *J. Colloid Interface Sci.* **2010**, *350*, 530–537. [CrossRef] [PubMed]

35. Song, X.H.; Yang, Y.F.; Liu, J.C.; Zhao, A.H.Y. PS colloidal particles stabilized by graphene oxide. *Langmuir* **2011**, *27*, 1186–1191. [CrossRef] [PubMed]

36. Zhang, X.; Wei, S.; Haldolaarachchige, N.; Colorado, H.; Luo, Z.; Young, D.P. Magnetoresistive conductive polyaniline–barium titanate nanocomposites with negative permittivity. *J. Phys. Chem. C* **2012**, *116*, 15731–15740. [CrossRef]

37. Gu, H.; Huang, Y.; Zhang, X.; Wang, Q.; Zhu, J.; Shao, L. Magnetoresistive polyaniline-magnetite nanocomposites with negative dielectrical properties. *Polymer* **2012**, *46*, 801–809. [CrossRef]

38. Zhu, J.; Gu, H.; Luo, Z.; Haldolaarachige, N.; Young, D.P.; Wei, S. Carbon nanostructure-derived polyaniline metacomposites: Electrical, dielectric, and giant magnetoresistive properties. *Langmuir* **2012**, *28*, 10246–10255. [CrossRef] [PubMed]

39. Huang, J.X.; Virji, S.; Weiller, B.H.; Kaner, R.B. Polyaniline nanofibers: Facile synthesis and chemical sensors. *J. Am. Chem. Soc.* **2003**, *125*, 314–315. [CrossRef] [PubMed]

40. Zhu, J.H.; Chen, M.J.; Qu, H.L.; Zhang, X.; Wei, H.G.; Luo, Z.P.; Colorado, H.A.; Wei, S.Y.; Guo, Z.H. Interfacial polymerized polyaniline/graphite oxide nanocomposites toward electrochemical energy storage. *Polymer* **2012**, *53*, 5953–5964. [CrossRef]

materials

MDPI

Communication

High-Surface-Area, Emulsion-Templated Carbon Foams by Activation of polyHIPEs Derived from Pickering Emulsions

Robert T. Woodward [1,*]**, François De Luca** [1]**, Aled D. Roberts** [2] **and Alexander Bismarck** [3]

[1] Department of Chemical Engineering, Imperial College London, South Kensington Campus, London SW7 2AZ, UK; f.de-luca13@imperial.ac.uk

[2] Department of Chemistry, University of Liverpool, Crown Street, Liverpool L69 7ZD, UK; aleddeakin@gmail.com

[3] Polymer and Composite Engineering (PaCE) Group, Institute of Materials Chemistry & Research, Faculty of Chemistry, University of Vienna, Währingerstraße 42, Vienna 1090, Austria; alexander.bismarck@univie.ac.at

* Correspondence: r.woodward@imperial.ac.uk; Tel.: +44-20-7589-5111

Academic Editors: To Ngai and Jonathan Phillips
Received: 30 June 2016; Accepted: 9 September 2016; Published: 14 September 2016

Abstract: Carbon foams displaying hierarchical porosity and excellent surface areas of >1400 m^2/g can be produced by the activation of macroporous poly(divinylbenzene). Poly(divinylbenzene) was synthesized from the polymerization of the continuous, but minority, phase of a simple high internal phase Pickering emulsion. By the addition of KOH, chemical activation of the materials is induced during carbonization, producing Pickering-emulsion-templated carbon foams, or carboHIPEs, with tailorable macropore diameters and surface areas almost triple that of those previously reported. The retention of the customizable, macroporous open-cell structure of the poly(divinylbenzene) precursor and the production of a large degree of microporosity during activation leads to tailorable carboHIPEs with excellent surface areas.

Keywords: carbon; polyHIPE; carboHIPE; Pickering emulsion; carbonization; microporous; emulsion template

---

1. Introduction

Carbonaceous materials possessing hierarchical porosity are desirable owing to a number of advantageous properties, one such example being the coupling of effective mass transfer through macropores and large surface areas derived from both meso- and micropores. High surface areas allow for a large degree of solid-liquid interactions, while effective mass transfer permits rapid access to this surface area throughout the material. More generally, carbonaceous materials are studied prolifically due to their relative chemical inertness and high thermal stability, which has resulted in their consideration as adsorbents [1] or catalyst supports [2], in gas sorption [3], and as electrodes in both Li-ion batteries [4] and double-layer capacitors [5].

The requirement for well-defined porosity within carbon structures has led to the utilization of a range of different precursor materials including carbides [6,7], synthetic organic polymers [8,9], and templated materials [10,11]. One attractive route to hierarchically porous carbon foams is via the carbonization of macroporous polymers templated from high internal phase emulsions (HIPEs). A HIPE is defined as an emulsion containing an internal phase that comprises >74.05% of the total volume of the system, the maximum volume that uniform spheres can close pack without deformation [12]. Polymers are produced from these templates by the polymerization of the continuous, but minority, phase of a HIPE to produce a polyHIPE. Aside from the influence of phase volume ratio and the emulsification method, a great degree of control over the structural properties of a polyHIPE's macropores is possible simply by varying the choice of emulsion stabilizer in the initial HIPE template. HIPEs are commonly stabilized using standard molecular surfactants, which upon polymerization produce open-cell polyHIPEs, meaning that the polymers contain pore throats which connect the emulsion-templated macropores, producing a more permeable structure. On the other hand, particle-stabilized emulsions, also known as Pickering emulsions, are stabilized by larger particles irreversibly adsorbing at the interface of emulsion [13,14]. Upon polymerization of Pickering HIPEs, closed-cell polyHIPEs containing no pore throats and macropores with larger diameters than those of the surfactant systems are produced [15,16]. However, it was demonstrated by Wong et al. that open-cell polyHIPEs could be produced from Pickering HIPEs via the addition of a small amount of a standard molecular surfactant prior to polymerization, introducing pore throats into Pickering-HIPE derived polyHIPEs [17].

When a charable polyHIPE is produced, it can be used as a precursor to porous carbon foams, or carboHIPEs, which retain the emulsion-templated macroporosity of the polyHIPE while developing micro- and mesopores upon carbonization, creating hierarchical porosity. The production of carboHIPEs has been described in the literature from a wide variety of starting materials, some of which include sulfonated poly(styrene-*co*-divinylbenzene) [18,19], Kraft black liquor [20], lignin [21], tannins [22], and polyacrylonitrile [23]. Recently, we described the production of carboHIPEs from poly(divinylbenzene)HIPEs synthesized from simple Pickering water-in-divinylbenzene (DVB) HIPE templates [24]. Carbonization of these templates gave carboHIPEs in good yields, retaining the emulsion-templated porosity of the poly(DVB)HIPE precursor and producing surface areas of up to 500 m^2/g. Pickering HIPEs have also been used to create carboHIPEs by producing sacrificial molds in the form of mesoporous macrocellular silica foams [25]. These Si(HIPE)s were infiltrated with a carbon precursor, carbonized, and washed in HF to produce carboHIPEs with surface areas of up to 900 m^2/g.

In order to try to improve the surface area of carbonaceous materials in general, varying methods of activation have been employed. One method involves the addition of activating agents to charable materials to try to increase the surface area of the resulting carbon by chemical activation during heating. Another involves the carbonization of materials in atmospheres containing small amounts of CO_2 or H_2O in order to induce physical activation of the materials. The literature is rich with studies on activation processes, with one of the most popular involving the addition of the chemical activating agent, KOH, to a material prior to carbonization [26]. Previous reports indicate that increases in porosity and surface area brought about by KOH-activation is due to a number of factors including chemical activation by etching of the carbon framework, physical activation by the gasification of the carbon, and carbon lattice expansion by the production of metallic K throughout the carbon framework [26–29]. It has been demonstrated that the activation of templated mesoporous carbons with well-controlled pore-size distributions led to huge increases in both surface area and total pore volumes by the production of micropores [30,31].

Herein, we report the synthesis of high-surface-area, hierarchically porous carboHIPEs by KOH-activation of poly(DVB) derived from Pickering HIPEs. The successful retention of emulsion-templated macroporosity during activation allows for carboHIPEs with almost triple the surface area of their non-activated (simply carbonized) equivalents. A combination of both silica particles and a small amount of standard molecular surfactant were used in order to create open-cell, charable structures. This increase in surface area, coupled with customizable macropore diameters, opens up opportunities to produce tailorable carboHIPEs with excellent surface areas.

2. Experimental

2.1. Materials

Calcium chloride (CaCl$_2$), divinylbenzene (DVB, 80% containing inhibitor), azobisisobutyronitrile (AIBN), and potassium hydroxide (KOH) were all purchased from Sigma-Aldrich (Dorset, UK) and used as received. Hydrophobic silica particles (HDK grade H20) were kindly provided by Wacker Chemie (Bracknell, UK), and Hypermer 2296 was kindly provided by Croda (Leek, UK). Both products were used as received. Silver paint used for SEM imaging was purchased from Agar Scientific (Stansted, UK).

2.2. Preparation of HIPEs and Subsequent poly(DVB)HIPEs

The synthesis of poly(DVB)HIPEs from Pickering water-in-DVB HIPEs has been described previously [24]. For a typical formulation, hydrophobic silica particles (120 mg, 3 wt %) were added to DVB (4 mL), and the sample was shaken by hand for ~5 min until the particles were well dispersed in the monomer. AIBN (1 mol % with respect to the monomer, 47 mg) was then added and the mixture stirred using a vortex mixture ((VortexGenie 2, Scientific Industries, Bohemia, NY, USA), speed setting '3', equating to roughly 600 rpm) before an aqueous CaCl$_2$ solution (10 g/L, 16 mL) was added slowly over 20 min while still stirring using the vortex mixer at the same speed setting. After the addition of the aqueous phase was complete, the HIPE was stirred more vigorously (speed setting '10', roughly 3000 rpm) for a further 5 min before the surfactant Hypermer 2296 (0.2 mL, 5 vol %) (Croda, UK) was added with respect to the initial monomeric phase, after which the HIPE was gently agitated by hand for 10 s. To create poly(DVB)HIPEs with smaller macropores, the HIPE was stirred at either 1000 rpm or 3000 rpm for 10 s at this point in the process, in place of gentle agitation by hand. HIPEs were then transferred into a 15 mL free standing polypropylene centrifuge (Falcon) tube (VWR, Radnor, PA, USA) and heated at 70 °C for 24 h in a convection oven to initiate polymerization. After polymerization, the Falcon tube containing the polyHIPE was cut into roughly 1 cm^3 cylinders using a band saw before the polyHIPE was removed from the tubes and washed three times in an ethanol bath for a total of at least 6 h. The cylinders were then dried in a vacuum oven at 110 °C overnight.

2.3. Preparation of carboHIPEs and Activated carboHIPEs

Samples of poly(DVB)HIPE were prepared for either carbonization or carbonization in the presence of a chemical activator. All samples were weighed before heating to 800 °C under a N$_2$ atmosphere at a ramp rate of 2 °C/min. Once 800 °C was reached, samples were held at this temperature for 1 h before the furnace was allowed to cool to room temperature overnight (remaining under a N$_2$ atmosphere). Samples being prepared for chemical activation were also weighed before being soaked in an aqueous KOH solution (either a 10 wt % or a 30 wt % solution), after which they were carefully removed and placed in a convection oven at 70 °C to dry overnight. After drying, samples were weighed again to determine how much KOH was deposited before being placed in the furnace and carbonized using the same process as the non-activated carboHIPEs. When investigating tailored average macropore diameters in activated carboHIPEs, all poly(DVB)HIPEs were immersed in a 30 wt % KOH solution before subsequent drying and carbonization as described above.

2.4. Characterization

Gas sorption analyses were performed on a Micromeritics 3Flex Surface Characterization Analyzer (Micromeritics, Atlanta, GA, USA) at −196 °C. Samples were degassed in situ under vacuum (around 0.0030 mbar) at 150 °C for at least 4 h, prior to measurement. SEM images were either taken on a variable pressure SEM (JEOL JSM 5610 LV (0.5–35 kV), Tokyo, Japan) or, in the case of the high-resolution SEM, images were taken using a high-resolution field emission gun SEM (FEGSEM (5 kV, InLens detector)) (Leo Gemini 1525 coupled with a SmartSEM software interface, Carl Zeiss NTS Ltd., Cambridge, UK). All polyHIPE samples were fixed on Al stubs (Agar Scientific Ltd., Stansted,

UK) using carbon tape to attach samples securely. The stubs were then sputtered with chromium (10 nm) and, taking care not to contaminate the sample, a small amount of silver DAG paint was used to provide a conductive bridge between the carbon tape and the Al stub. Image analysis, such as measuring the average macropore diameter of polyHIPEs, was carried out using the image software ImageJ (version 1.48, National Institutes of Health, Bethesda, MD, USA) [32]. The percentage porosity (P) was determined from both the envelope (ρ_e) and skeletal density (ρ_s) via the equation P = $(1 - \rho_e/\rho_s) \times 100\%$.

2.5. Research and Discussion

Water-in-DVB emulsions were prepared using a combination of silica particles and standard molecular surfactants as stabilizers. The use of the Pickering emulsifiers was crucial as we previously demonstrated that poly(DVB)HIPEs derived from solely surfactant-stabilized HIPEs did not survive carbonization [24]. After curing the emulsions for 24 h at 70 °C, all white poly(DVB)HIPEs were produced (Figure 1a), which displayed emulsion-templated macroporosity with an average diameter of 82 μm (Table 1), in a similar range to other polyHIPEs produced from Pickering HIPEs [16,33]. As a mixed surfactant system was used, pore throats were created in the resulting poly(DVB)HIPE, producing the desired open-cell structure (Figure 1b). Poly(DVB)HIPEs were carbonized at 800 °C in an inert N_2 atmosphere and yielded carbon foams, or carboHIPEs (Figure 1a). The carboHIPEs retained the cylindrical shape of poly(DVB)HIPE precursors well, albeit with a significant volume loss of between 73% and 76%. The emulsion-templated macropores also survived carbonization to yield an open-cell carboHIPE, showing a decrease in their average diameter of 20 μm to 62 μm. (Figure 1c).

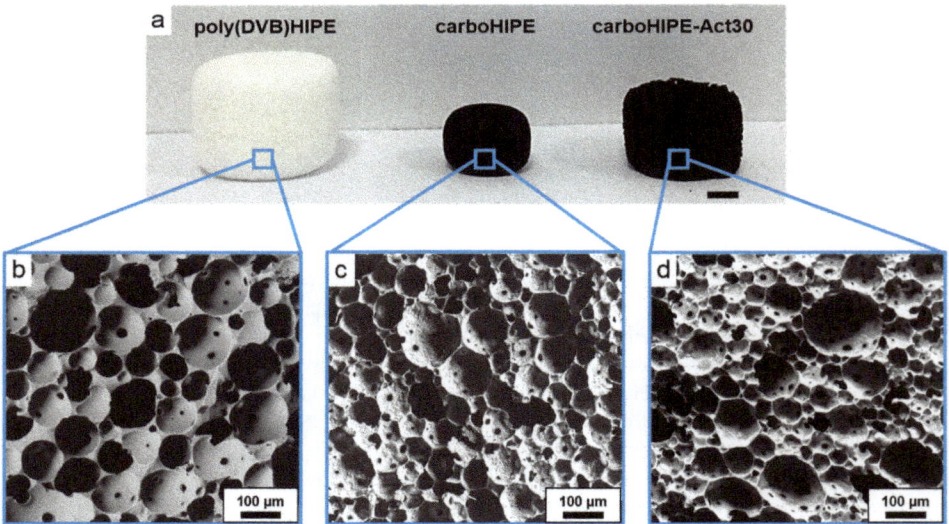

Figure 1. (**a**) Photograph of a poly(DVB)HIPE, carboHIPE, and a carboHIPE-Act30; (**b–d**) SEM images of a poly(DVB)HIPE, carboHIPE, and a carboHIPE-Act30, respectively. The scale bar in the photograph represents 2 mm.

Table 1. Macropore diameter, BET surface area, micropore volume, total pore volume, porosity, and char yield of a poly(DVB)HIPE, a carboHIPE, and activated carboHIPEs.

Sample	Average Macropore Diameter (μm) [a]	Surface Area (m²/g) [b]	Micropore vol. (cm³/g) [b]	Total Pore vol. (g/cm³) [b]	Porosity (%) [c]	Char Yield (%) [d]
poly(DVB)HIPE	82 ± 42	8	0	0.021	86	N/A
carboHIPE	62 ± 28	521	0.268	0.223	95	22
carboHIPE-Act10	72 ± 26	1123	0.432	0.572	97	13
carboHIPE-Act30	74 ± 30	1456	0.554	0.791	97	12

[a] Measured using image analysis software; [b] Calculated from N_2 sorption isotherms at 77 K; [c] Calculated using both the skeletal and envelope density of monoliths and [d] Mass yield after carbonization, relative to original poly(DVB)HIPE.

Varying levels of activation of poly(DVB)HIPEs were achieved by submerging poly(DVB)HIPEs in KOH solution of varying concentrations for two hours under gentle agitation, prior to carbonization. The poly(DVB)HIPEs were then dried in an oven and carbonized in a similar procedure to the non-activated carboHIPEs. Activated materials will be denoted by the concentration of the KOH solution in which they were submerged; for example, in carboHIPE-Act10, the 'Act' refers to activation and '10' indicates that the poly(DVB)HIPE precursor was exposed to a 10 wt % KOH solution. Activation of poly(DVB)HIPEs appeared to lead to slightly larger carboHIPEs than simple carbonization, with volume losses in activated materials ranging between 50% and 61% (Figure 1a), although this was admittedly hard to quantify in many samples due to the more irregular shapes and slightly more brittle nature of activated materials. The higher retention in volume may be due to the increased evolution of gas exerting outward pressure on structures during activation, or intercalated K in the carbon framework preventing further collapse. The emulsion-templated macropores were successfully retained during activation (Figure 1d) and showed a slightly reduced average macropore diameter of 74 μm for carboHIPE-Act30, in good agreement with the reduced volume loss in comparison to simple carbonization in the absence of KOH. In order to demonstrate the tailoring of macropore size in activated carboHIPEs, standard water-in-DVB HIPEs were stirred at varied rates after the addition of the Hypermer 2296 surfactant, prior to curing and activation. The samples were either gently agitated by hand (as is the case in the standard procedure), or stirred for 10 s at 1000 rpm or 3000 rpm using a vortex mixer. Average macropore diameters of 74 ± 30 μm, 18 ± 5 μm, and 11 ± 4 μm were obtained for the standard carboHIPE-Act30, and those stirred at 1000 and 3000 rpm, respectively (Figure 2), demonstrating controllable macropore size in activated carboHIPEs.

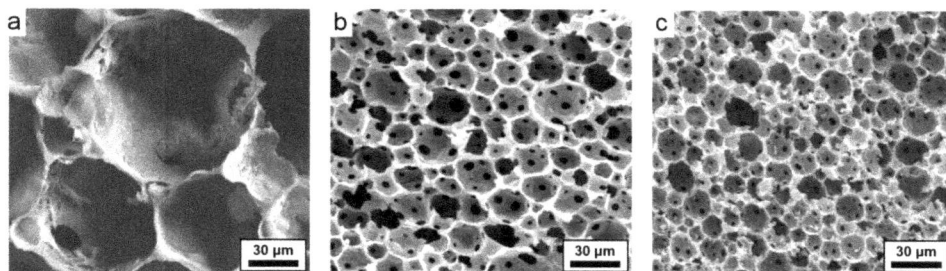

Figure 2. SEM images of various carboHIPE-Act30s with different average macropore diameters, produced by stirring the HIPEs at different speeds prior to polymerization. (**a**) Gently agitated by hand; (**b**) 1000 rpm for 10 s on a vortex mixer; and (**c**) 3000 rpm for 10 s on a vortex mixer.

Gas sorption analysis was performed on samples, applying the BET model for surface area measurements and determining the micropore volume using the t-plot method. The native poly(DVB)HIPE had a low surface area of 8 m^2/g due to the presence of some mesoporosity within the structure [24,34]. All surface areas showed huge increases upon carbonization or activation, with nitrogen adsorption displaying a type I isotherm in all cases (Figure 3a), with a steep N$_2$ uptake at low pressures, indicative of a predominantly microporous structure, as outlined by IUPAC classification [35]. The surface areas of carboHIPEs increased significantly with the addition of KOH prior to carbonization (Table 1), producing excellent surface areas of up to 1456 m^2/g for carboHIPE-Act30. Both the total pore volume and the micropore volume also improved dramatically upon carbonization or activation, increasing from no notable microporosity in the poly(DVB)HIPE to an excellent micropore volume of 0.544 cm^3/g and a total pore volume of 0.791 cm^3/g in carboHIPE-Act30. Pore size distributions show this increase in micropore volume with a large peak at ~5 Å after both carbonization and activation, with activated samples appearing significantly more microporous (Figure 3b). By increasing the concentration of the initial KOH solution used to deposit the activating agent, both the surface area and the micropore volume could significantly increase (Figure 3, Table 1) without any significant structural damage occurring to the emulsion-templated macropores (Figure 1, Table 1).

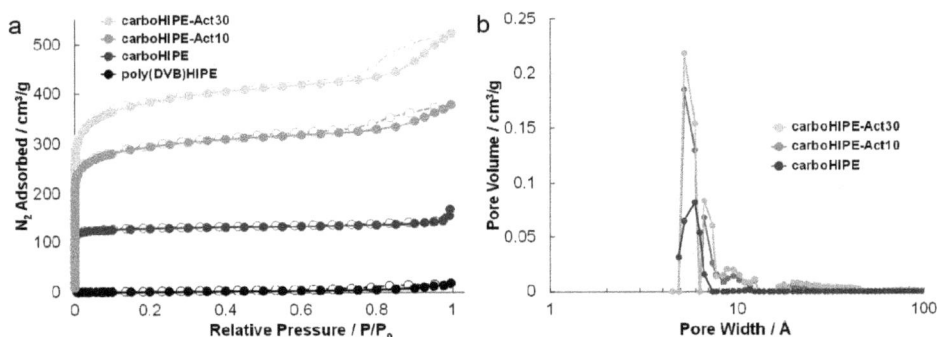

Figure 3. (a) N$_2$ adsorption/desorption isotherms of all materials including a poly(DVB)HIPE, a carboHIPE, and carboHIPEs with varying degrees of activation (carboHIPE-Act10 and carboHIPE-Act30). Filled shapes show adsorption while empty shapes show desorption; (b) Non-local density functional theory pore size distribution for a carboHIPE, carboHIPE-Act10, and carboHIPE-Act30.

In order to further investigate the effect of the activating agent on the emulsion-templated structure and what may lead to the large increases in surface area, high magnification SEM was performed. When no activation agent was used during carbonization, the surface of the carboHIPE looked relatively smooth and unbroken with the exception of the pore throats in the images (Figure 4a,b). However, in carboHIPE-Act30, the surface was cracked and appeared more damaged after activation in comparison to carbonization with no KOH (Figure 4c,d). Considering the high magnification images of both the carboHIPE and carboHIPE-Act30, a much more porous structure is observed for the latter (Figure 4), with the visible pores mainly in the macropore region (>50 nm in diameter). This newly formed porous structure may allow better access to the increased amount of micropores throughout the structure, resulting in much higher surface areas.

Lastly, the degree of graphitization between samples was probed using Raman spectroscopy (Figure 5). It is clear from the spectra that carbonization/activation led to the formation of D and G modes (1350 cm^{-1} and 1582 cm^{-1}, respectively) and the loss of the typical fine structure of the poly(DVB)HIPE in the Raman spectrum [36]. The formation of these broad D and G peaks is indicative of disordered graphitic carbonaceous structures [37], typical for carbonization at relatively low temperatures [38]. The D to G peak intensity ratio is around 0.9 for all samples, suggesting that the degree of graphitization is independent of the degree of activation in the carboHIPEs.

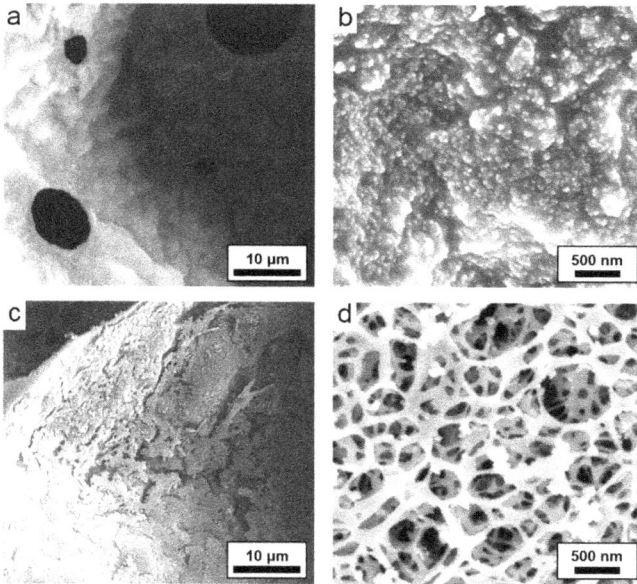

Figure 4. SEM images. (**a**,**b**) show a typical non-activated carboHIPE; and (**c**,**d**) show a typical activated carboHIPE-Act30.

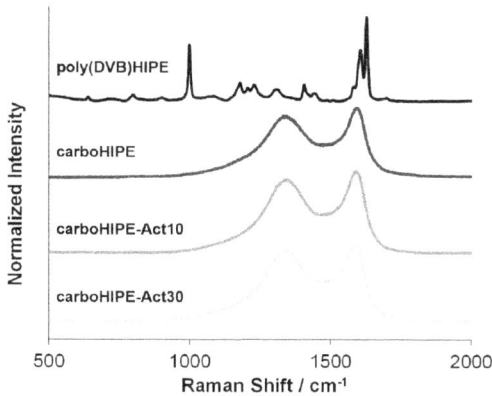

Figure 5. Raman spectroscopy of a poly(DVB)HIPE, a carboHIPE, and activated carboHIPEs.

3. Conclusions

By simple immersion and subsequent drying of Pickering-emulsion derived poly(DVB)HIPEs in inexpensive KOH solutions, carboHIPEs containing hierarchical porosity and surface areas of up to 1456 m^2/g were easily produced upon carbonization. This addition of this simple step almost triples both the resulting surface area and micropore volume of carboHIPEs in comparison to poly(DVB)HIPEs carbonized in the presence of no activating agent. The degree of microporosity was controlled by varying the amount of KOH present during carbonization. The activation process was not detrimental to the emulsion-templated porosity of the poly(DVB)HIPEs, and it was demonstrated that the size of the macropores could also be dictated in activated carboHIPEs by controlling the amount of energy input after the addition of the standard surfactant. The retention of the macropores along with the production of large micropore volumes opens up potential for efficient mass transfer of electrolytes to the increased surface areas of these carbon foams, meaning they may have potential as novel monolithic electrodes in supercapacitor devices and will be investigated further. The tunable porosity of these materials on both the macro- and the micropore scale could also lead to their use as efficient adsorbents for the removal of organic pollutants, or as high surface area catalyst supports.

Acknowledgments: We wish to thank Ryan T. Luebke and Camille Petit (Imperial College London) for assistance with the gas sorption measurements shown. We also wish to thank David B. Anthony (Imperial College London) for assistance with Raman spectroscopy. We gratefully acknowledge funding from the UK Engineering and Physical Sciences Research Council (EPSRC) through the projects "graphene 3D networks" (EP/K01658X/1) and "nanoporous materials" (EP/J014974/1). The authors declare no conflict of interest.

Author Contributions: Robert T. Woodward and Alexander Bismarck conceived and designed the experiments; Robert T. Woodward performed the experiments and analysis; François De Luca contributed the high magnification SEM analysis; Aled D. Roberts devised and optimized the activation method and Robert T. Woodward wrote the paper.

Conflicts of Interest: The authors declare no conflict of interest.

References

1. Mohan, D.; Sarswat, A.; Ok, Y.S.; Pittman, C.U.J. Organic and inorganic contaminants removal from water with biochar, a renewable, low cost and sustainable adsorbent—A critical review. *Bioresour. Technol.* **2014**, *160*, 191–202. [CrossRef] [PubMed]
2. Wildgoose, G.G.; Banks, C.E.; Compton, R.G. Metal nanopartictes and related materials supported on carbon nanotubes: Methods and applications. *Small* **2006**, *2*, 182–193. [CrossRef] [PubMed]
3. Kemp, K.C.; Seema, H.; Saleh, M.; Le, N.H.; Mahesh, K.; Chandra, V.; Kim, K.S. Environmental applications using graphene composites: Water remediation and gas adsorption. *Nanoscale* **2013**, *5*, 3149–3171. [CrossRef] [PubMed]
4. Roberts, A.D.; Li, X.; Zhang, H.F. Porous carbon spheres and monoliths: Morphology control, pore size tuning and their applications as Li-ion battery anode materials. *Chem. Soc. Rev.* **2014**, *43*, 4341–4356. [CrossRef] [PubMed]
5. Sevilla, M.; Mokaya, R. Energy storage applications of activated carbons: Supercapacitors and hydrogen storage. *Energy Environ. Sci.* **2014**, *7*, 1250–1280. [CrossRef]
6. Gogotsi, Y.; Nikitin, A.; Ye, H.H.; Zhou, W.; Fischer, J.E.; Yi, B.; Foley, H.C.; Barsoum, M.W. Nanoporous carbide-derived carbon with tunable pore size. *Nat. Mater.* **2003**, *2*, 591–594. [CrossRef] [PubMed]
7. Yushin, G.; Dash, R.; Jagiello, J.; Fischer, J.E.; Gogotsi, Y. Carbide-derived carbons: Effect of pore size on hydrogen uptake and heat of adsorption. *Adv. Funct. Mater.* **2006**, *16*, 2288–2293. [CrossRef]
8. Meng, Y.; Gu, D.; Zhang, F.Q.; Shi, Y.F.; Cheng, L.; Feng, D.; Wu, Z.; Chen, Z.; Wan, Y.; Stein, A.; et al. A family of highly ordered mesoporous polymer resin and carbon structures from organic-organic self-assembly. *Chem. Mater.* **2006**, *18*, 4447–4464. [CrossRef]
9. Bear, J.C.; McGettrick, J.D.; Parkin, I.P.; Dunnill, C.W.; Hasell, T. Porous carbons from inverse vulcanized polymers. *Microporous Mesoporous Mater.* **2016**, *232*, 189–195. [CrossRef]
10. Yoon, S.B.; Kim, J.Y.; Yu, J.S. Synthesis of highly ordered nanoporous carbon molecular sieves from silylated MCM-48 using divinylbenzene as precursor. *Chem. Commun.* **2001**. [CrossRef]
11. Lu, A.H.; Schuth, F. Nanocasting: A versatile strategy for creating nanostructured porous materials. *Adv. Mater.* **2006**, *18*, 1793–1805. [CrossRef]
12. Cameron, N.R.; Sherrington, D.C. High internal phase emulsions (HIPEs)—Structure, properties and use in polymer preparation. In *Biopolymers Liquid Crystalline Polymers Phase Emulsion*; Springer: Berlin/Heidelberg, Germany, 1996; pp. 163–214.
13. Pickering, S.U. CXCVI.—Emulsions. *J. Chem. Soc. Trans.* **1907**, *91*, 2001–2021. [CrossRef]
14. Binks, B.P.; Lumsdon, S.O. Influence of particle wettability on the type and stability of surfactant-free emulsions. *Langmuir* **2000**, *16*, 8622–8631. [CrossRef]

15. Menner, A.; Verdejo, R.; Shaffer, M.; Bismarck, A. Particle-stabilized surfactant-free medium internal phase emulsions as templates for porous nanocomposite materials: Poly-pickering-foams. *Langmuir* **2007**, *23*, 2398–2403. [CrossRef] [PubMed]

16. Ikem, V.O.; Menner, A.; Bismarck, A. High Internal Phase Emulsions Stabilized Solely by Functionalized Silica Particles. *Angew. Chem. Int. Ed.* **2008**, *47*, 8277–8279. [CrossRef] [PubMed]

17. Wong, L.L.C.; Ikem, V.O.; Menner, A.; Bismarck, A. Macroporous Polymers with Hierarchical Pore Structure from Emulsion Templates Stabilised by Both Particles and Surfactants. *Macromol. Rapid Commun.* **2011**, *32*, 1563–1568. [CrossRef] [PubMed]

18. Wang, D.; Smith, N.L.; Budd, P.M. Polymerization and carbonization of high internal phase emulsions. *Polym. Int.* **2005**, *54*, 297–303. [CrossRef]

19. Asfaw, H.D.; Younesi, R.; Valvo, M.; Maibach, J.; Ångström, J.; Tai, C.-W.; Bacsik, Z.; Sahlberg, M.; Nyholm, L.; Edström, K. Boosting the thermal stability of emulsion–templated polymers via sulfonation: An efficient synthetic route to hierarchically porous carbon foams. *Chem. Sel.* **2016**, *1*, 784–792. [CrossRef]

20. Foulet, A.; Birot, M.; Backov, R.; Sonnemann, G.; Deleuze, H. Preparation of hierarchical porous carbonaceous foams from Kraft black liquor. *Mater. Today Commun.* **2016**, *7*, 108–116. [CrossRef]

21. Ungureanu, S.; Sigaud, G.; Vignoles, G.L.; Lorrette, C.; Birot, M.; Deleuze, H.; Backov, R. First Biosourced Monolithic Macroporous SiC/C Composite Foams (Bio-SiC/C(HIPE)) Bearing Unprecedented Heat Transport Properties. *Adv. Eng. Mater.* **2013**, *15*, 893–902. [CrossRef]

22. Szczurek, A.; Fierro, V.; Pizzi, A.; Celzard, A. Emulsion-templated porous carbon monoliths derived from tannins. *Carbon* **2014**, *74*, 352–362. [CrossRef]

23. Cohen, N.; Silverstein, M.S. Synthesis of emulsion-templated porous polyacrylonitrile and its pyrolysis to porous carbon monoliths. *Polymer* **2011**, *52*, 282–287. [CrossRef]

24. Woodward, R.T.; Fam, D.W.H.; Anthony, D.B.; Hong, J.; McDonald, T.O.; Petit, C.; Shaffer, M.S.P.; Bismarck, A. Hierarchically porous carbon foams from pickering high internal phase emulsions. *Carbon* **2016**, *101*, 253–260. [CrossRef]

25. Ungureanu, S.; Birot, M.; Deleuze, H.; Schmitt, V.; Mano, N.; Backov, R. Triple hierarchical micro-meso-macroporous carbonaceous foams bearing highly monodisperse macroporosity. *Carbon* **2015**, *91*, 311–320. [CrossRef]

26. Wang, J.C.; Kaskel, S. KOH activation of carbon-based materials for energy storage. *J. Mater. Chem.* **2012**, *22*, 23710–23725. [CrossRef]

27. Romanos, J.; Beckner, M.; Rash, T.; Firlej, L.; Kuchta, B.; Yu, P.; Suppes, G.; Wexler, C.; Pfeifer, P. Nanospace engineering of KOH activated carbon. *Nanotechnology* **2012**, *23*. [CrossRef] [PubMed]

28. Wang, H.L.; Gao, Q.M.; Hu, J. High Hydrogen Storage Capacity of Porous Carbons Prepared by Using Activated Carbon. *J. Am. Chem. Soc.* **2009**, *131*, 7016–7022. [CrossRef] [PubMed]

29. Lozano-Castello, D.; Calo, J.M.; Cazorla-Amoros, D.; Linares-Solano, A. Carbon activation with KOH as explored by temperature programmed techniques, and the effects of hydrogen. *Carbon* **2007**, *45*, 2529–2536. [CrossRef]

30. Jin, J.; Tanaka, S.; Egashira, Y.; Nishiyama, N. KOH activation of ordered mesoporous carbons prepared by a soft-templating method and their enhanced electrochemical properties. *Carbon* **2010**, *48*, 1985–1989. [CrossRef]

31. Xing, W.; Huang, C.C.; Zhuo, S.P.; Yuan, X.; Wang, G.Q.; Hulicova-Jurcakova, D.; Yan, Z.F.; Lu, G.Q. Hierarchical porous carbons with high performance for supercapacitor electrodes. *Carbon* **2009**, *47*, 1715–1722. [CrossRef]

32. Schneider, C.A.; Rasband, W.S.; Eliceiri, K.W. NIH Image to ImageJ: 25 years of image analysis. *Nat. Methods* **2012**, *9*, 671–675. [CrossRef] [PubMed]

33. Menner, A.; Ikem, V.; Salgueiro, M.; Shaffer, M.S.P.; Bismarck, A. High internal phase emulsion templates solely stabilised by functionalised titania nanoparticles. *Chem. Commun.* **2007**, *43*, 4274–4276. [CrossRef]

34. Hainey, P.; Huxham, I.M.; Rowatt, B.; Sherrington, D.C.; Tetley, L. Synthesis and ultrastructural studies of styrene divinylbenzene polyhipe polymers. *Macromolecules* **1991**, *24*, 117–121. [CrossRef]

35. Sing, K.S.W.; Everett, D.H.; Haul, R.A.W.; Moscou, L.; Pierotti, R.A.; Rouquerol, J.; Siemieniewska, T. Reporting physisorption data for gas solid systems with special reference to the determination of surface-area and porosity (recommendations 1984). *Pure Appl. Chem.* **1985**, *57*, 603–619. [CrossRef]

36. Stokr, J.; Schneider, B.; Frydrychova, A.; Coupek, J. Composition analysis of crosslinked styrene-ethylene dimethacrylate and styrene-divinylbenzene copolymers by raman-spectroscopy. *J. Appl. Polym. Sci.* **1979**, *23*, 3553–3561. [CrossRef]

37. Ferrari, A.C.; Robertson, J. Interpretation of Raman spectra of disordered and amorphous carbon. *Phys. Rev. B* **2000**, *61*, 14095–14107. [CrossRef]

38. Li, J.; Su, S.; Zhou, L.; Abbot, A.M.; Ye, H. Dielectric transition of polyacrylonitrile derived carbon nanofibers. *Mater. Res. Express* **2014**, *1*. [CrossRef]

materials

MDPI

Article

Pickering Particles Prepared from Food Waste

Joanne Gould [1,*], Guillermo Garcia-Garcia [2] and Bettina Wolf [1]

[1] Division of Food Sciences, School of Biosciences, The University of Nottingham,
    Sutton Bonington Campus, Loughborough LE12 5RD, UK; bettina.wolf@nottingham.ac.uk
[2] Centre for SMART, Wolfson School of Mechanical and Manufacturing Engineering,
    Loughborough University, Loughborough LE11 3TU, UK; G.Garcia-Garcia@lboro.ac.uk
*   Correspondence: joanne.gould@nottingham.ac.uk; Tel.: +44-115-951-6134

Academic Editors: To Ngai and Syuji Fujii
Received: 9 August 2016; Accepted: 14 September 2016; Published: 21 September 2016

Abstract: In this paper, we demonstrate the functionality and functionalisation of waste particles as an emulsifier for oil-in-water (o/w) and water-in-oil (w/o) emulsions. Ground coffee waste was chosen as a candidate waste material due to its naturally high content of lignin, a chemical component imparting emulsifying ability. The waste coffee particles readily stabilised o/w emulsions and following hydrothermal treatment adapted from the bioenergy field they also stabilised w/o emulsions. The hydrothermal treatment relocated the lignin component of the cell walls within the coffee particles onto the particle surface thereby increasing the surface hydrophobicity of the particles as demonstrated by an emulsion assay. Emulsion droplet sizes were comparable to those found in processed foods in the case of hydrophilic waste coffee particles stabilizing o/w emulsions. These emulsions were stable against coalescence for at least 12 weeks, flocculated but stable against coalescence in shear and stable to pasteurisation conditions (10 min at 80 °C). Emulsion droplet size was also insensitive to pH of the aqueous phase during preparation (pH 3–pH 9). Stable against coalescence, the water droplets in w/o emulsions prepared with hydrothermally treated waste coffee particles were considerably larger and microscopic examination showed evidence of arrested coalescence indicative of particle jamming at the surface of the emulsion droplets. Refinement of the hydrothermal treatment and broadening out to other lignin-rich plant or plant based food waste material are promising routes to bring closer the development of commercially relevant lignin based food Pickering particles applicable to emulsion based processed foods ranging from fat continuous spreads and fillings to salad dressings.

Keywords: pickering emulsions; particles; lignin; food emulsions

---

1. Introduction

Pickering particles are solid particles capable of stabilising an emulsion by the adsorption of solid particles to the oil/water interface. The application of Pickering particles has attracted significant research interest in recent years as unlike molecular emulsifiers, which constantly adsorb and desorb from the interface promoting emulsion droplet coalescence, Pickering particles are considered to be irreversibly adsorbed. This is because the free energy needed for spontaneous desorption of particles from the interface is extremely large compared to that of thermal energy. For example, for desorption of a particle of radius 10 nm adsorbed at a toluene-water interface with a contact angle of 90° the energy required is 2750 KT [1]. The particle is therefore considered to be permanently adsorbed as the high desorption energy means a high energy input is needed to disrupt the particle layers to allow droplet coalescence to occur. This holds true for all particle stabilised emulsions even for small nanoparticles ($r \approx 5$–10 nm) as long as the contact angle of the particle is not too close to 0° or 180° [2].

The properties of particle stabilised emulsions (droplet size, flocculation, viscosity) are majorly influenced by the properties of the particles and emulsion phases controlling the arrangement of the particles at the interface. Particle wettability is a key determinant of whether an oil-in-water (o/w) or a water-in-oil (w/o) emulsion is obtained, commonly characterised by the contact angle at the interface measured through the water phase. Particles classed as hydrophilic adopt a contact angle of less than 90° at a planar air/water or oil/water interface, i.e., these are preferentially wetted by water. Conversely, particles forming contact angles of greater than 90° are hydrophobic and are wetted by the oil phase to a greater extent [3,4]. During emulsion formation, the interface of a droplet will curve to ensure the larger area of the particle surface remains on the external side, such that hydrophilic particles will give rise to o/w emulsions and hydrophobic particles to w/o emulsions [3,4].

Utilisation of Pickering particles in emulsion based foods and other consumer good products, e.g., creams and lotions, offers several advantages such as replacement of artificial surfactants, prolonged shelf life, and stabilisation of complex structures such as multiple emulsions. However, the inclusion of these particles in food products is hampered through the lack of interfacially active food particles. Hydrophobic OSA modified starches [5], flavonoids [6], chitin nanocrystals extracted from crab shells [7], fat particles such as hardened rapeseed oil particles [8], protein based particles [9], protein microgels [10], egg yolk granules [11], and colloidal cellulose based fibers [12] have all been shown to have interfacial functionality although most often chemical modification is required before use. We have, on the other hand, recently demonstrated that particles from the shell and nib of the Theobroma Cacao pod act as Pickering particles. O/W emulsions readily formed during high shear mixing processes showed no evidence of a change in emulsion droplet size over 100 days of storage or the presence of an oil layer after storage for two years, indicating the formation of a highly stable microstructure [13,14]. These particles are not only food grade but also natural as there is no requirement for chemical modification. Further investigations of these natural Pickering particles indicated that the emulsifying ability of the particles was enhanced by the presence of lignin.

Lignin is the second most abundant natural polymer after cellulose, characterised by its highly branched heterogeneous structure built from aromatic residues. It is widely considered to be a hydrophobic molecule, however it has also been shown to have hydrophilic, hydrophobic, and amphiphilic character depending on botanical origin and extraction methods [15,16]. Kraft lignins [17,18], lignosulfonates [18], lignin obtained from enzymatic hydrolysis [19] and lignin microparticles [20,21] have been shown to stabilise o/w emulsions. Several methods have been used to create lignin microparticles; one such method uses aqueous ethanol to extract lignin from shrub willow and an anti-solvent precipitation protocol to prepare the microparticles. The emulsifying ability of the lignin microparticles was then assessed in a soybean oil-in-water system with the result being the formation of stable o/w emulsions with no significant change in droplet size over a storage period of five months [20].

To the best of our knowledge there have yet to be published reports on the preparation of lignin based Pickering particles for the application in w/o emulsions warranting microstructure stability. We hypothesise that a lignin rich particulate material can be suitably processed to show functionality in this system, a functionality that we were not able to impart to cocoa particles. Their lignin content of between 4% and 9% (wt/wt) [22] is either too low or the modification methods explored in this research are not suitable. Here we selected ground coffee waste with a reported lignin content of between 20% and 27% (wt/wt) [23]. It is in plentiful supply with UK coffee shops and households producing more than 500,000 tonnes [24] and 60,379 tonnes [25] of this type of waste per year respectively. However, we foresee the waste produced during the manufacture of instant coffee, termed spent coffee grounds, also to be a suitable waste stream. Application of this technology to spent coffee grounds provides an even larger commercial potential as from the 2.5 million tonnes of coffee products manufactured in Europe in 2013, 326,320 tonnes correspond to instant coffee, with a monetary value close to €3 billion [26]. This large scale production generates over 300,000 tonnes of dry spent coffee grounds every year [27].

In order to prepare these Pickering particles we have utilised a hydrothermal treatment common to the bioenergy industry with the aim of relocating the lignin to the surface of the waste coffee particles. In doing so, we expect to increase the hydrophobicity of the particle allowing stabilisation of w/o emulsion in addition to the untreated particles stabilizing o/w emulsions. The hydrothermal treatment is carried out in water at high temperatures and pressures which causes the cell wall to be disrupted and the formation of spherical droplets on the surface of the material which are understood to be largely composed of lignin [28]. Although, lignin is notoriously complex to characterize [29] multiple techniques such as FT-IR, NMR, antibody labelling, TEM, and cytochemical staining have been used to investigate the composition of the droplets, all of which determined the droplets to be composed of lignin [30,31]. The formation of droplets occurs because, at temperatures above the melting point of lignin, typically between 100 °C and 170 °C [32], lignin fluidizes, coalesces, and has the ability to move through the cell wall matrix. Once at the surface of the sample material, the hydrophobic lignin minimizes contact with the hydrophilic solvent by forming droplets which solidify once the temperature has been brought down sufficiently [32,33]. However, some authors do conclude that, during steam explosion and dilute acid treatments. Hemicellulose and lignin degradation products combine to form lignin like droplets termed pseudo-lignin [34]. Pseudo-lignin is also considered to be hydrophobic like lignin [31,35].

In this paper, we demonstrate that particles prepared from ground coffee waste show promise to be successfully applied as a versatile Pickering emulsifier through hydrothermal processing. To the best of our knowledge, this is the first report of a natural non-fat based Pickering particle suitable for application as an emulsifier of w/o food emulsions. Data presented include size, hydrophobicity, and emulsifying ability for untreated and treated ground waste coffee particles. For the sake of brevity, ground coffee waste particles are in the following referred to as coffee particles or simply particles.

2. Results and Discussion

2.1. Properties of Prepared Coffee Pickering Particles

2.1.1. Surface Morphology

The surface morphology of a coffee particle following drying after collection (Figure 1A), that was then milled (B) and additionally submitted to hydrothermal treatment at various temperatures (Figure 1C–F) as assessed using SEM is depicted in Figure 1. The untreated coffee particle shown in Figure 1A is characterised by an irregular surface with folded rounded features. Ball milling of these particles had little effect on the surface morphology of the particles except a slight smoothing of the surface as shown in Figure 1B.

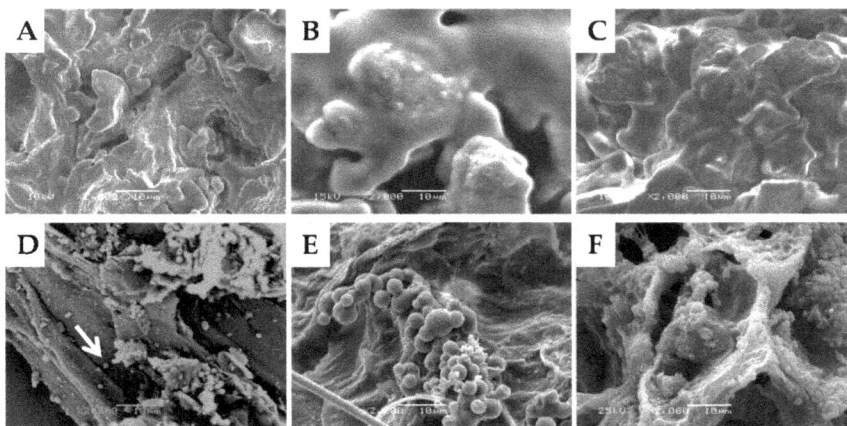

Figure 1. SEM images of (**A**) dried coffee particle; (**B**) dried and ball milled coffee particle and (**C–F**) following hydrothermal treatment at (**C**) 150 °C; (**D**) 200 °C; (**E**) 250 °C; and (**F**) 275 °C for 1 h. Scale bar represents 10 μm.

Treating the coffee particles hydrothermally at temperatures between 150 °C and 275 °C for 1 h caused the formation of droplets on the particles' surface, which was a result of the melting and relocation of the lignin component, as discussed in the introduction. By relocating the lignin in this manner, we predict the surface hydrophobicity of the coffee particles will increase enabling the stabilisation of w/o emulsions. The effect of temperature was therefore investigated to optimize the formation of these droplets on the surface, as it has previously been reported that the density of coalesced lignin located on the cell structure of hydrothermally treated sugarcane bagasse increased with temperature [36].

Figure 1D–F demonstrate that the droplets on the surface of the coffee particles appear in clusters. Clustering around and within specific structural features such as pits, cell corners, and delamination layers has previously been reported after hydrothermal treatment of corn stem rind and explained by the porosity of these areas [30]. The pores act as extrusion channels for the melted lignin. If lignin droplets were to be found more evenly distributed across the particle surface, as has been reported for hydrothermally treated sugarcane bagasse, this would be the sign of a porous ultrastructure which may be generated in situ during hydrothermal processing due to the removal and hydrolysis of lignin and hemicellulose, respectively [36].

The amount of droplets formed on the particle surface as judged by the SEM images depended on the temperature of the hydrothermal treatment (Figure 1C–F). While the SEM image 1(C) acquired after treatment at 150 °C featured no droplets, Figure 1D indicates that hydrothermal treatment at 200 °C led to the formation of a small number of small droplets with diameters of around 1 μm. Evidenced in Figure 1E, more droplets and varying in size between 2.5 μm and 5 μm formed following treatment at 250 °C. Some droplets appear to have fused together. Also recognisable are flattened edges where a droplet may have been in close contact with another droplet that was subsequently pulled off during preparation of the sample for imaging. Similar droplet features were found on the surface of corn stover rind following hydrothermal and acid treatment [30]. Figure 1F then demonstrates that further increase in temperature is detrimental to the occurrence of surface adsorbed lignin droplets. This observation suggests that there is an optimal processing temperature to impart surface hydrophobicity to coffee particles should this indeed be the functionality of what is assumed to be mostly redistributed lignin. Therefore, the sample treated at 250 °C for 1 h was selected for further investigations.

2.1.2. Lignin Content

The concentration of lignin in the particles prepared from ground coffee waste was quantified spectrophotometrically following acetyl bromide and dioxane extraction. Lignin content was 17.9% ± 1.2% in milled particles and 29.9% ± 1.2% in hydrothermally treated milled particles (250 °C for 1 h). Lignin was not created nor lost during the hydrothermal process, instead the increase in lignin content is an effect of sample mass loss due to the hydrolysis of the indigenous polysaccharides xylan and hemicellulose during hydrothermal treatment [37,38]. Such an increase has previously been reported for wheat straw following steam explosion treatment [39].

2.1.3. Particle Size

The particle size of the coffee Pickering particles is shown in Table 1 as the volume weighted mean diameter, $d_{4,3}$, and as a measure of the fine fraction the diameter below which 10% of the particles are found, $d_{10,3}$. Ball milling decreased the values of both characteristic diameters. The hydrothermal treatment appeared to slightly increase in the mean diameter which would be the result of particle aggregation during the processing. Hydrothermal treatment did not affect the size of the fine fraction which is worth noting since our previous research on the emulsifying ability of cocoa particles has shown that the size of the fine fraction dictated the diameter of the o/w emulsions processed with these particles [13]. For a new Pickering particle system as under investigation here one would base any expectation on emulsion droplet size on the mean particle diameter. The linear relationship between emulsion droplet diameter and particle diameter stabilising the interface [40] suggests that the milled and hydrothermally treated particles would stabilise smaller emulsion droplets than the unmilled coffee particles.

Table 1. Volume-weighted characteristic particle sizes of coffee particles examined for Pickering properties. All particles were dried prior to particle size analysis.

Sample	$d_{4,3}$ (μm)	$d_{10,3}$ (μm)
Unmilled	341.11 ± 21.67	35.91 ± 5.65
Milled	117.12 ± 16.15	21.90 ± 2.55
Milled and Hydrothermally Treated	144.25 ± 6.14	19.24 ± 0.20

2.1.4. Particle Hydrophobicity

The hydrophobicity of the Pickering particles was evaluated using an emulsion assay, as measuring the contact angle, the material property commonly chosen to characterise the hydrophobicity of Pickering particles—of particles with irregular shapes—can introduce significant errors, depending on the method of assessment. Instead, the emulsion composition and processing protocol described in 3.3 was designed to give an insight into the hydrophobicity of the particles, based on the type of emulsion formed.

Unmilled and milled coffee particles stabilised o/w emulsions regardless of the oil phase composition (polarity) which was confirmed with the drop test where all emulsions dispersed in water rather than in a sample of the oil phase of the emulsion. The stabilisation of o/w emulsions by these Pickering particles indicates their hydrophilic nature, as even with the most polar oil phase (100% IPM) o/w emulsions were formed.

Figure 2 shows the microstructures of a selection of the o/w emulsions stabilised by unmilled (Figure 2A–C) and milled coffee particles (Figure 2D–F). The micrographs show that the Pickering emulsions have a flocculated microstructure and the milling of the particles enabled the stabilisation of smaller droplets, in accordance with the smaller particle size of the milled sample reported in Table 1. Oil droplets stabilised by unmilled coffee particles showed a broad size distribution which makes differentiating the effect of oil polarity on emulsification efficacy difficult. In contrast, the size of emulsion droplets stabilised by milled coffee particles was affected by the oil phase polarity, with larger droplets stabilised when the oil phase consisted of equal quantities of IPM and dodecane as evident in Figure 2E. In absence of experimental evidence, we suggest that the altered interfacial properties or viscosity properties of the oil phase may be the reason for the formation of the larger droplets.

In contrast, emulsions processed in the presence of hydrothermally treated coffee particles formed w/o emulsions regardless of the polarity of the oil phase, again confirmed by the result of the drop test where all emulsions dispersed in the oil phase. This result demonstrates that hydrothermal treatment increased the hydrophobicity of the particles most likely due to the relocation of lignin to the particle surface.

Figure 3 shows the microstructures of the different w/o emulsions (varying oil phase polarity) stabilised by milled and hydrothermally treated coffee particles. Comparison of Figure 3 to Figure 2 reveals obvious microstructure differences. The droplet size in the w/o emulsions was considerably larger than the droplet size in the o/w emulsions featuring droplets with diameter of between 100 μm and 300 μm. Another difference is the occurrence of irregular shaped water droplets, pointed out by the arrows in Figure 3B–E, as an intermediate stage of coalescence—termed arrested coalescence—in which droplets retain the shape of the original droplets to some extent. Complete coalescence is halted when the interface is jammed with particles preventing further reduction in interfacial area to a spherical droplet. This phenomenon is therefore strongly dependent on the level of droplet surface coverage with particles [41]. A high degree of droplet surface coverage with particles ($\varnothing = 0.9$) creates a closely packed jammed interfacial layer preventing total coalescence. In the emulsions shown in Figure 3, the droplets are described as in state of arrested coalescence. At lower surface coverage and if the combined particle covered surface area of two droplets exceeds the interfacial area that would form by complete coalescence arrested coalescence will occur. Based on experimental data, it was deducted that a combined intermediate particle surface coverage of the two droplets of $1.43 < \varnothing_1 + \varnothing_2 < 1.81$ is required to prevent total coalescence, leaving droplets in an arrested state of coalescence [41]. The presence of irregular shaped water droplets in the case of hydrothermally treated coffee Pickering particles therefore indicates an intermediate level of water droplet surface coverage with these coffee particles.

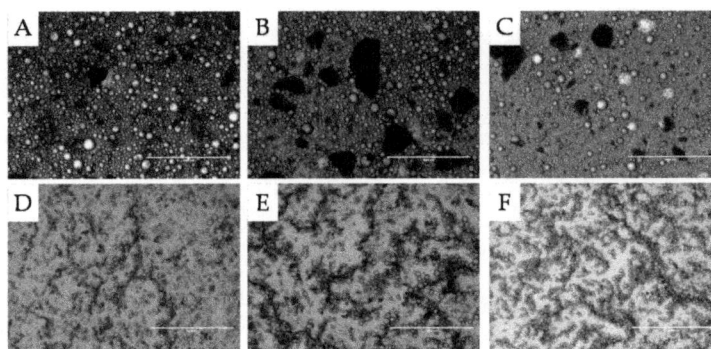

Figure 2. Light micrographs of o/w emulsions with 46% oil and stabilised with 8% unmilled or milled coffee particles with different oil phase polarity acquired after one day of storage at 25 °C. Images are as follows; unmilled coffee particle with an oil phase of (**A**) 100% dodecane (least polar); (**B**) 50% dodecane and 50% IPM (intermediate polarity) and (**C**) 100% IPM (most polar) and milled coffee particles (**D**) 100% dodecane; (**E**) 50% dodecane and 50% IPM and (**F**) 100% IPM. Scale bar represents 1000 µm.

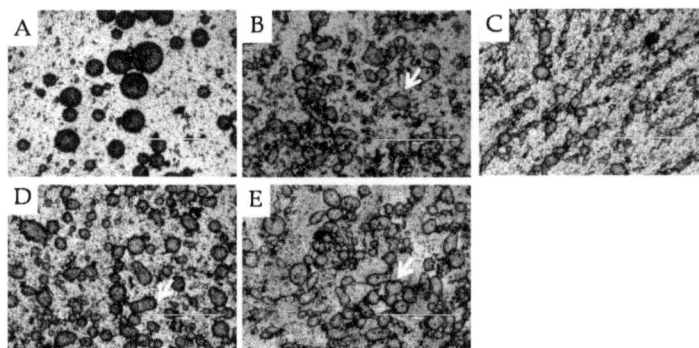

Figure 3. Light micrographs of w/o emulsions with 46% water and stabilised with 8% hydrothermally treated coffee particles with differing oil phase composition acquired after one day of storage at 25 °C. Images are of w/o emulsions with oil phases of (**A**) 100% dodecane; (**B**) 75% dodecane; (**C**) 50% dodecane; (**D**) 25% dodecane; and (**E**) 100% IPM. Scale bar represents 1000 µm. The arrows indicate the presence of arrested coalescence.

This intermediate surface coverage may be the result of the heterogeneous distribution of lignin droplets on the particle surface evidenced in Figure 1. These overall hydrophobic Pickering particles are characterised by inhomogeneous surface chemistry, and adsorption onto the water droplet surface appears to be possible only for the more hydrophilic or lignin-poor surface domains. Particle adsorption therefore maybe characterised by large parts of the particles residing in the water phase resulting in a lower surface coverage and causing the larger size of the stabilised droplets compared to the o/w emulsions with untreated particles of a most likely relatively homogenous surface chemistry. Optimisation of the hydrothermal treatment may allow a more homogenous or control over surface chemistry/hydrophobicity allowing an increase in particle adsorption to the interface and therefore an increased surface coverage by particles, preventing arrested coalescence of emulsion droplets.

2.2. Application of Coffee Pickering Particles in Food Emulsions

The processability of o/w emulsions stabilised with milled waste coffee Pickering particles was evaluated to see whether coffee Pickering particles could be successfully incorporated into manufactured food products. The processability and storage stability of these Pickering emulsions was evaluated in a food emulsion formulation consisting of 46% sunflower oil and by subjecting the emulsions to shearing and heating as well as acidic and alkaline conditions. W/O emulsions were not characterised in terms of processability as the large size of the water droplets stabilised by hydrothermally treated waste coffee particles are not currently desirable in food products due to their predictable instability towards shear and mixing processes as well as potentially imparting a rough mouthfeel as the water droplets may be sensed as large solid particles due to the arrested nature of the interface.

2.2.1. Microstructure Stability

Figure 4 shows the size distribution of o/w emulsions stabilised by milled waste coffee particles measured after 1 day, 6 weeks, and 12 weeks of storage at 25 °C, alongside the size distribution of an aqueous dispersion of the milled waste coffee particles. The emulsion had a bimodal distribution with sharp peaks at 100 μm and 500 μm. It is also evident that there was a significant overlap between the size distribution of the emulsion droplets and the particles that stabilise the emulsion droplets. Based on the microscopy evidence shown in Figure 5A the peak at 100 μm can be assigned to the emulsion droplets. Due to conclusions from previous literature [2], we expect the particles that stabilise the emulsion droplets to be in the order of one magnitude smaller in diameter which would correspond to the distribution of particles below 10 μm. Particle sizes around 500 μm identify particle aggregates that can be noted in Figure 5A or individual very large particles having a large impact on particle size distribution due to weighting by volume. It is also worth noting that the difference between the emulsion and dispersion distributions between 10 μm and 30 μm is indicative of the presence of unadsorbed suspended particles in the emulsion.

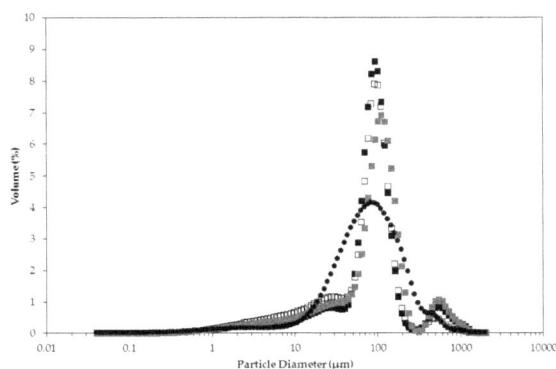

Figure 4. Emulsion droplet diameter volume size distribution of o/w emulsions stabilised with 8% milled waste coffee Pickering particles and sunflower oil as the oil phase (46%) acquired after 1 day (■), 6 weeks (□), and 12 weeks (■) of storage at 25 °C, is presented alongside the size distribution of the particles (●) Data presented is the mean distribution of three independent emulsion samples.

As shown in Figure 4, the peak at 100 μm remained constant over the studied storage period of 12 weeks and there was no significant change in the mean diameter over storage as can be seen in Figure 7. There were minor changes in the volume fraction of the larger particles (peak around 500 μm) which we expect to be due to sampling of the large particle aggregates or individual large particles.

2.2.2. Temperature Stability

To ensure products are safe for consumption high temperature processing steps such as pasteurisation, sterilisation, and cooking are often used in food manufacturing. It was therefore important to investigate the influence of heating and holding the emulsion at 80 °C for 10 min on the emulsion microstructure. Figure 5A,B presents the microstructure before and after heating, respectively. There is little difference in the degree of flocculation and droplet size evident, which was reflected in the results of the particle size analysis on these two samples (data not presented). The heat stability of the coffee particle Pickering system indicates that these emulsions could be utilised in thermally processed foods.

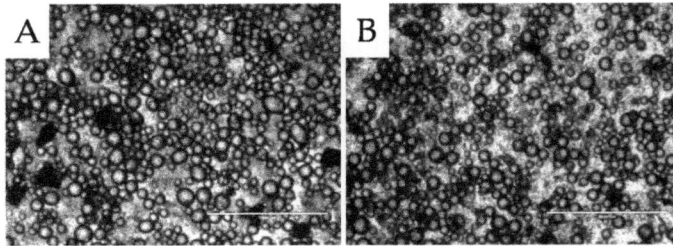

Figure 5. Light micrographs of o/w emulsions stabilised with 8% milled waste coffee Pickering particles and sunflower oil as the oil phase (46%) before heating (**A**); and after heating (**B**) the emulsion to 80 °C. Scale bar represents 1000 μm.

2.2.3. Shear Stability

In addition to storage and temperature stability, the microstructure of the coffee particle stabilised emulsions showed good shear stability as shown in Figure 6. The viscosity data were acquired by increasing and then decreasing the shear rate which was repeated three times in total. The emulsion is clearly shear thinning and the slight shoulder between $1 \ s^{-1}$ and $10 \ s^{-1}$ indicates that some slip occurred. Nevertheless, following the first attainment of the highest shear rate, the viscosity data overlapped at each shear rate and the viscosity recorded at the highest shear rate was constant independent of the step in the measurement sequence. This behaviour is highly indicative of the equilibration of the emulsion's superstructure during the first shear rate increase in response to the shear rate applied, thus viscosity subsequently probed at a shear rate lower or equal to this maximum shear rate remained unchanged. There was no evidence of droplet break up caused by the shearing protocol as the mean size, characterised by the $d_{4,3}$, of the emulsion droplets was not significantly ($p < 0.05$) different in emulsions before and after shearing (data not shown). Typically, it is flocculation of emulsion droplets and in this case potentially also of aggregates of non-adsorbed coffee particles that are broken up during the first shear rate increase in such up and down shear rate protocols. Indeed, the emulsion as shown in Figure 5A appears slightly flocculated.

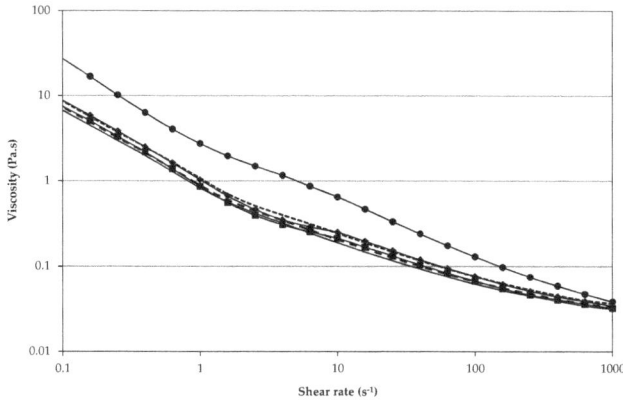

Figure 6. Shear stability was evaluated by applying a shear rate increase to 1000 s^{-1} and decrease to 0 s^{-1}, the protocol was repeated a total of three times. Black symbols indicate shear increase steps (1st●, 2nd♦, and 3rd■) and straight lines represent shear decrease steps (1st········, 2nd— — —, and 3rd———).

2.2.4. pH Stability of Milled Waste Coffee Pickering Emulsions

Application in manufactured foods will expose the emulsions to a range of ingredients including acids and alkalis. In order to assess whether these changes could cause emulsion destabilisation, aqueous phases were adjusted to pH 3, 6, and 9 prior to homogenisation and the stability of the emulsions formed were assessed. Figure 7 shows the volume mean emulsion droplet diameters of the pH adjusted emulsions after one day and four weeks of storage alongside data acquired on an emulsion formed with pure water as the aqueous phase. Altering the pH of the aqueous phase between 3 and 9 did not have a significant effect on the mean emulsion droplet size over the storage period.

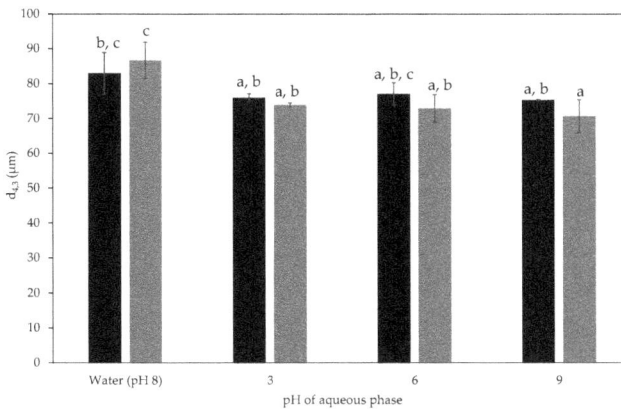

Figure 7. Volume mean emulsion droplet diameter acquired one day (■) and four weeks (■) after emulsification. The emulsions were formed with aqueous phases adjusted to pH 3, 6, 9. The droplet size of emulsions formed with pure water (pH 8) as the aqueous phase was included this data was acquired after one day (■) and six weeks (■). Emulsion formulation contained 8% milled waste coffee Pickering particles and 46% sunflower oil. The presence of different letters (a,b,c) represent a significant difference between samples ($p < 0.05$).

3. Materials and Methods

3.1. Materials

Ground coffee waste produced from a variety of ground coffee products was collected from a local coffee outlet. Sodium azide (Sigma-Aldrich, Dorset, UK) was added as antimicrobial agent to all aqueous emulsion phases to give a final concentration of 0.02% w/w. Double distilled water was used for all samples. The oil phase of the emulsions varied in composition of isopropyl myristate (IPM) and dodecane (Sigma-Aldrich, Dorset, UK) and commercially available sunflower oil (purchased at a local supermarket). Acetyl bromide (Sigma-Aldrich, Dorset, UK), glacial acetic acid (Fisher Scientific, Loughborough, UK), and low sulfonate kraft lignin (Sigma-Aldrich, Dorset, UK) were used to quantify the lignin content of the milled ground coffee waste particles and hydrothermally treated ground coffee waste particles. Hydrochloric acid (SG 1.16, 32%) (HCl) (Fisher Scientific, Loughborough, UK) and sodium hydroxide pellets (NaOH) (Fisher Scientific, Loughborough, UK) were used to adjust the pH of the aqueous phase to pH 3, 6, and 9. All of these materials were used as received.

3.2. Prepartation of Ground Coffee Waste Particle Preparation

Pickering particles were prepared from this material after drying in a convection oven at 40 °C for 48 h to a moisture content of 7.3% ± 0.2% and milling by dry grinding in a planetary ball mill (PULVERISETTE 5 classic line, Fritsch GmbH, Oberstein, Germany) for particle size reduction. The milling conditions were 10 zirconium oxide (ZrO_2) balls 15 mm in diameter and 70 g of ground coffee waste particles in a 500 mL ZrO_2 grinding bowl at a main disc speed setting of 200 rpm for 12 h. The milling programme consisted of 5 min intervals of milling and no milling to minimize heat production. Hydrothermally treated milled ground coffee waste particles were prepared by hydrothermally treating the milled waste coffee particles using a protocol adapted from literature [24]. 4 g of sample was mixed with 40 ml of water and sealed into stainless steel tubular reactors (17 cm long and 3 cm inner diameter). Loaded reactors were held at selected temperatures between 150 °C and 275 °C for 1 h. At the conclusion of the treatment, the tubes were cooled by submerging in cold water for 5 min. Solids were retained by filtration (Whatman Grade 1, Kent, UK) and dried in a convection oven at 40 °C for 48 h to a final moisture content of 5.5% ± 0.1%.

3.3. Emulsion Processing

All emulsions regardless of the components were prepared as follows. Emulsions were produced on a 50 g scale, containing 46% oil, 46% double distilled water, and 8% particles, based on preliminary experiments (data not presented) where it was found that emulsions containing less than 8% particles were unstable to coalescence. Particles were added as a powder on top of the water phase (the densest liquid phase) followed by the oil phase, in accordance with the powdered particle method [42] as this removes any effect of the initial location of the particles on their wettability which could influence the type of emulsion formed (o/w or w/o). The mixture was emulsified using a high shear overhead mixer (L5M Series fitted with emulsor screen, Silverson, Chesham, UK) operating at 9000 rpm for 2 min. The emulsion type was confirmed by observing whether a drop of the emulsion dispersed in pure oil or in pure water, with a w/o emulsion dispersing in oil and an o/w emulsion dispersing in water [42].

3.4. Characterisation of Ground Coffee Waste Pickering Particles and Pickering Emulsions

3.4.1. Lignin Quantification

The milled ground coffee waste and hydrothermally treated (250 °C, 1 h) coffee waste particles' lignin content was quantified by initially extracting the lignin using acetyl bromide followed by measurement of absorbance (UV-Vis Spectrophotometer, Varian Cary 50, Agilent Technologies, Santa Clara, CA, USA) at 280 nm [24]. Briefly, 100 mg of sample material was dissolved in 4 mL of solvent (25% acetyl bromide, 75% glacial acetic acid) followed by incubation at 50 °C for 2 h. Quantification was performed by calibration using the low sulphonate kraft lignin as a reference material.

3.4.2. Microstructure Imaging

The surface structure of the untreated, milled and hydrothermally treated ground coffee waste particles was investigated using SEM (JSM 6060LV, JEOL, Tokyo, Japan). The particles were placed on SEM stubs with carbotape followed by drying under vacuum before being coated in gold using gold splutter (Leica SCD 0005, Leica Microsystems, Milton Keynes, UK). The samples were then transferred to the SEM stage for imaging.

The microstructure of emulsions was visualised utilizing bright field microscopy (EVOS f1, AMG, Washington, DC, USA) with the aim to support particle sizing data for water continuous emulsions. In the case of oil-continuous emulsions, bright field microscopy was the only method applied to get an insight into droplet size.

3.4.3. Particle Sizing

Size distributions for the aqueous suspensions of waste coffee particles and oil-in-water emulsions were acquired with a low angle laser diffraction particle size analyser (LS 13 320, Beckman Coulter, High Wycombe, UK) fitted with a dispersion cell filled with water (Universal liquid module, LS13 320, Beckman Coulter, High Wycombe, UK). Three independent replicates of each sample were taken. Data was analysed using the Fraunhofer diffraction model using the instrument's software. Oil-continuous emulsions were not analysed in this equipment due to the reported shear sensitivity of these emulsions and potential droplet size reduction due to the mixing and pumping processes in the dispersion cell.

3.4.4. Particle Hydrophobicity

Particle hydrophobicity was evaluated through an emulsion assay where the lipophilic emulsion phase is varied in polarity and water is the hydrophilic emulsion phase. Following the emulsion preparation method described in 3.3 removing volume fraction of either emulsion phase and initial location of particle addition as impacting factors, the type of emulsion formed provides an indication of particle hydrophobicity as the particles will transfer into the continuous emulsion phase during emulsification, with hydrophilic particles stabilising an o/w emulsion and hydrophobic particles stabilising a w/o emulsion. The polarity of the oil phase was altered to contain varying quantities (0%, 25%, 50%, 75%, and 100%) of IPM, a polar oil, and dodecane, a non-polar oil, as it has been shown in emulsions stabilised with silica particles of intermediate hydrophobicity the nature of the oil can affect the final type of the emulsion formed. Binks et al. [43] found that a polar oil e.g., IPM, interacts more strongly with water; the strength of the interaction can be quantified in terms of the work of adhesion and is calculated from the interfacial tension of the system. As the adhesion between the phases increases, due to an increase in polarity, the water fraction at which phase inversion also increases, therefore for the most polar oils the work of adhesion is too high so no phase inversion occurs allowing only w/o emulsions to be formed whereas non-polar oils preferentially form o/w emulsions. However, as we will show, the hydrophobicity of the Pickering particles can overrule this affect.

3.4.5. Process Stability of o/w Emulsions

To evaluate whether o/w emulsions stabilised by milled ground coffee waste Pickering particles could be successfully incorporated into manufactured food products, sunflower oil–in–water emulsions stabilised with milled ground coffee waste particles were subjected to shear and heat as well as acidic and alkaline conditions as relevant to process steps such as mixing, pumping, pasteurisation, and pH adjustment to achieve desired product textures or to contribute to microbial stability of the product.

The shear stability of the emulsions was evaluated at 20 °C using a rotational rheometer (MCR301, Anton Paar, Graz, Austria) fitted with a concentric cylinder geometry (bob diameter: 27 mm, bob length: 40 mm, cup diameter: 29 mm) applying the following shear ramp protocol. The shear rate was increased from 0.1 to 1000 s^{-1} in 5 min followed by a shear rate decrease from 1000 to 0.1 s^{-1} in 5 min. The shear ramp was repeated three times with the viscosity recorded and plotted against shear rate for all shear ramps.

The influence of temperature on emulsion stability was examined by placing the emulsions in individual glass vials followed by incubation in a water bath at 80 °C for 10 min. The pH of the aqueous phases of the emulsions were adjusted to pH 3, 6, and 9 by adding either HCl or NaOH prior to emulsification. The stability of emulsions subjected to high temperature or different pH environments was evaluated by assessing changes to droplet size and microstructure of the emulsions. Due to the presence of particle aggregates and potentially individual large particles, the particle size data were manipulated to remove any data at sizes greater than 400 μm, therefore only changes to the emulsion droplet size could be assessed.

3.5. Statistical Analysis

Mean size and standard deviation are reported based on three independent samples. The particle size data was significantly analysed using an ANOVA and Tukey's statistical test with the level of significance set at $p = 0.05$ for both statistical tests.

4. Conclusions

Waste coffee particles (unmilled and milled) and hydrothermally treated waste coffee particles can act as Pickering particles for both o/w and w/o emulsions. The Pickering stabilisation is a result of the presence and, for w/o application the relocation of the known emulsifying agent, lignin. Pickering particle hydrophobicity assessment confirmed that the relocation of the lignin to the particle surface had increased the hydrophobicity of the particles compared to the hydrophilic untreated particles. However, emulsions stabilised with the hydrothermally treated waste coffee Pickering particles had large droplets which would not be suitable for incorporation into food products, necessitating further research into process optimisation for this application.

On the other hand, the use of milled waste coffee Pickering particles to stabilise o/w emulsions of typical food formulation produced emulsion droplets of desirable size with no change in microstructure seen over a period of 12 weeks of storage. Investigations also showed that the o/w emulsions to be stable to shearing up to 1000 s^{-1} and heating to 80 °C, conditions typical of food product manufacture. Finally, altering the pH of the aqueous phase to between pH 3 and pH 9 was not found not to affect the stability of the emulsions with no change in droplet size seen for a period of four weeks.

Overall, this study has demonstrated that lignin rich food waste can be functionalised as a food ingredient with emulsifying property for both oil and water based foods. While application in water based foods, i.e., for o/w emulsions, appears to be more readily possible and is thus potentially closer to application, further research is required to develop commercially relevant particles for the application in lipid continuous foods or for the stabilisation of the encapsulated water phase in duplex (w/o/w) food emulsions. In addition, the natural abundance of lignin in plant based materials and plant based food waste begs to extend this application to materials other than coffee. Due to the complex and variable structure of lignin, the use of different sources could enable the creation of particles with a range of functionalities and applications.

Acknowledgments: This research was funded through the EPSRC Centre for Innovative Manufacturing in Food grant (EP/K030957/1). We would like to thank Christine Grainger-Boultby, Khat Husain, and Roger Ibbett for technical support.

Author Contributions: J.G. and B.W. conceived the research project and designed experiments. J.G. carried out all experiments. G.G.G. reviewed the literature on waste coffee. J.G. drafted the manuscript; B.W. supervised its preparation; all authors were involved in revising it. All authors have approved and are accountable for the final version of the manuscript submitted.

References

1. Binks, B.P. Particles as surfactants-similarities and differences. *Curr. Opin. Coll. Interface Sci.* **2002**, *7*, 21–41. [CrossRef]
2. Dickinson, E. Use of nanoparticles and microparticles in the formation and stabilization of food emulsions. *Trends Food Sci. Technol.* **2012**, *24*, 4–12. [CrossRef]
3. Aveyard, R.; Binks, B.P.; Clint, J.H. Emulsions stabilised solely by colloidal particles. *Adv. Colloid Interface Sci.* **2003**, *100*, 503–546. [CrossRef]
4. Finkle, P.; Draper, H.D.; Hildebrand, J.H. The theory of emulsification. *J. Am. Chem. Soc.* **1923**, *45*, 2780–2788. [CrossRef]
5. Yusoff, A.; Murray, B.S. Modified starch granules as particle-stabilizers of oil-in-water emulsions. *Food Hydrocoll.* **2011**, *25*, 42–55. [CrossRef]
6. Luo, Z.J.; Murray, B.S.; Yusoff, A.; Morgan, M.R.A.; Povey, M.J.W.; Day, A.J. Particle-stabilizing effects of flavonoids at the oil-water interface. *J. Agric. Food Chem.* **2011**, *59*, 2636–2645. [CrossRef] [PubMed]
7. Tzoumaki, M.V.; Moschakis, T.; Kiosseoglou, V.; Biliaderis, C.G. Oil-in-water emulsions stabilized by chitin nanocrystal particles. *Food Hydrocoll.* **2011**, *25*, 1521–1529. [CrossRef]
8. Paunov, V.N.; Paunov, V.N.; Cayre, O.; Nobel, P.F.; Stoyanov, S.D.; Velikov, K.P.; Golding, M. Emulsions stabilised by food colloid particles: Role of particle adsorption and wettability at the liquid interface. *J. Colloid Interface Sci.* **2007**, *312*, 381–389. [CrossRef] [PubMed]
9. De Folter, J.W.J.; van Ruijven, M.W.M.; Velikov, K.P. Oil-in-water Pickering emulsions stabilized by colloidal particles from the water-insoluble protein zein. *Soft Matter* **2012**, *8*, 2807–2815. [CrossRef]
10. Destribats, M.; Rouvet, M.; Gehin-delval, C.; Schmitt, C.; Binks, B.P. Emulsions stabilised by whey protein microgel particles: Towards food-grade Pickering emulsions. *Soft Matter* **2014**, *10*, 6941–6954. [CrossRef] [PubMed]
11. Rayner, M.; Marku, D.; Eriksson, M.; Sjöö, M.; Dejmek, P.; Wahlgren, M. Biomass-based particles for the formulation of Pickering type emulsions in food and topical applications. *Colloid Surf. A Physicochem. Eng. Appl.* **2014**, *458*, 48–62. [CrossRef]
12. Campbell, A.L.; Holt, B.L.; Stoyanov, S.D.; Paunov, V.N. Scalable fabrication of anisotropic micro-rods from food-grade materials using an in shear flow dispersion-solvent attrition technique. *J. Mater. Chem.* **2008**, *18*, 4074–4078. [CrossRef]
13. Gould, J.; Vieira, J.; Wolf, B. Cocoa particles for food emulsion stabilisation. *Food Funct.* **2013**, *4*, 1369–1375. [CrossRef] [PubMed]
14. Vieira, J.B.; Husson, J.; Wolf, B.; Gould, J. Emulsion Stabilisation. Patent EP2589297, 2013.
15. Doherty, W.O.; Mousavioun, P.; Fellows, C.M. Value-adding to cellulosic ethanol: Lignin polymers. *Ind. Crops Prod.* **2011**, *33*, 259–276. [CrossRef]
16. Tortora, M.; Cavalieri, F.; Mosesso, P.; Ciaffardini, F.; Melone, F.; Crestini, C. Ultrasound driven assembly of lignin into microcapsules for storage and delivery of hydrophobic molecules. *Biomacromolecules* **2014**, *15*, 1634–1643. [CrossRef] [PubMed]
17. Afanas'ev, N.; Selyanina, S.; Selivanova, N. Stabilization of the oleic acid-water emulsion with various kraft lignins. *Russ. J. Appl. Chem.* **2008**, *81*, 1851–1855. [CrossRef]
18. Gundersen, S.A.; Sæther, Ø.; Sjöblom, J. Salt effects on lignosulfonate and Kraft lignin stabilized O/W-emulsions studied by means of electrical conductivity and video-enhanced microscopy. *Colloid Surf. A Physicochem. Eng. Asp.* **2001**, *186*, 141–153. [CrossRef]
19. Gan, M.; Pan, J.; Zhang, Y.; Dai, X.; Yin, Y.; Qu, Q.; Yan, Y. Molecularly imprinted polymers derived from lignin-based Pickering emulsions and their selectively adsorption of lambda-cyhalothrin. *Chem. Eng. J.* **2014**, *257*, 317–327. [CrossRef]
20. Stewart, H.E. Development of Food-Grade Microparticles from Lignin. Ph.D. Thesis, Massey University, Palmerston North, New Zealand, 2015.
21. Wei, Z.; Yang, Y.; Yang, R.; Wang, C. Alkaline lignin extracted from furfural residues for pH-responsive Pickering emulsions and their recyclable polymerization. *Green Chem.* **2012**, *14*, 3230–3236. [CrossRef]
22. Redgwell, R.J.; Trovato, V.; Curti, D. Cocoa bean carbohydrates: Roasting-induced changes and polymer interactions. *Food Chem.* **2003**, *80*, 511–516. [CrossRef]

23. Pujol, D.; Gominho, J.; Olivella, M.A.; Fiol, N.; Villaescusa, I.; Pereira, H. The chemical composition of exhausted coffee waste. *Ind. Crops Prod.* **2013**, *50*, 423–429. [CrossRef]

24. Business Waste: How to Keep Coffee Grounds Out of the Ground. Avaliable online: http://www.gd-environmental.co.uk/blog/business-waste-how-keep-coffee-grounds-out-ground/ (accessed on 1 June 2016).

25. Chapagain, A.; James, K. *The Water and Carbon Footprint of Household Food and Drink Waste in the UK*; WRAP and WWF: Banbury, UK, 2011.

26. *European Coffee Report 2013/2014: European Chapter and Key National Data*; European Coffee Federation: Brussels, Belgium, 2014.

27. Obruca, S.; Benesova, P.; Kucera, D.; Petrik, S.; Marova, I. Biotechnological conversion of spent coffee grounds into polyhydroxyalkanoates and carotenoids. *New Biotechnol.* **2015**, *32*, 569–574. [CrossRef] [PubMed]

28. Pu, Y.Q.; Hu, F.; Huang, F.; Ragauskas, A.J. Lignin Structural Alterations in Thermochemical Pretreatments with Limited Delignification. *Bioenergy Res.* **2015**, *8*, 992–1003. [CrossRef]

29. Dean, J.F. Lignin analysis. In *Methods in Plant Biochemistry and Molecular Biology*, 1st ed.; CRC Press: Cleveland, OH, USA, 1997; pp. 199–215.

30. Donohoe, B.S.; Decker, S.R.; Tuvker, M.P.; Himmel, M.E.; Vinzant, T.B. Visualizing Lignin Coalescence and Migration Through Maize Cell Walls Following Thermochemical Pretreatment. *Biotechnol. Bioeng.* **2008**, *101*, 913–925. [CrossRef] [PubMed]

31. Araya, F.; Troncoso, E.; Mendonca, R.T.; Freer, J. Condensed lignin structures and re-localization achieved at high severities in autohydrolysis of Eucalyptus globulus wood and their relationship with cellulose accessibility. *Biotechnol. Bioeng.* **2015**, *112*, 1783–1791. [CrossRef] [PubMed]

32. Ma, J.; Zhang, X.; Zhou, X.; Xu, F. Revealing the Changes in Topochemical Characteristics of Poplar Cell Wall During Hydrothermal Pretreatment. *BioEnergy Res.* **2014**, *7*, 1358–1368. [CrossRef]

33. Ma, X.J.; Cao, S.L.; Luo, X.L.; Chen, L.H.; Huang, L.L. Surface characterizations of bamboo substrates treated by hot water extraction. *Bioresour. Technol.* **2013**, *136*, 757–760. [CrossRef] [PubMed]

34. Sannigrahi, P.; Kim, D.H.; Jung, S.; Ragauskas, A. Pseudo-lignin and pretreatment chemistry. *Energy Environ. Sci.* **2011**, *4*, 1306–1310. [CrossRef]

35. Hu, F.; Jung, S.; Ragauskas, A. Impact of Pseudolignin versus Dilute Acid-Pretreated Lignin on Enzymatic Hydrolysis of Cellulose. *ACS Sustain. Chem. Eng.* **2013**, *1*, 62–65. [CrossRef]

36. Reddy, P.; Lekha, P.; Reynolds, W.; Kirsch, C. Structural characterisation of pretreated solids from flow-through liquid hot water treatment of sugarcane bagasse in a fixed-bed reactor. *Bioresour. Technol.* **2015**, *183*, 259–261. [CrossRef] [PubMed]

37. Ibbett, R.; Gaddipati, S.; Greetham, D.; Hill, S.; Tucker, G. The kinetics of inhibitor production resulting from hydrothermal deconstruction of wheat straw studied using a pressurised microwave reactor. *Biotechnol. Biofuels* **2014**, *7*, 1. [CrossRef] [PubMed]

38. Ko, J.K.; Kim, Y.; Ximenes, E.; Ladisch, M.R. Effect of liquid hot water pretreatment severity on properties of hardwood lignin and enzymatic hydrolysis of cellulose. *Biotechnol. Bioeng.* **2015**, *112*, 252–262. [CrossRef] [PubMed]

39. Heiss-Blanquet, S.; Zheng, D.; Ferreira, N.L.; Lapierre, C.; Baumberger, S. Effect of pretreatment and enzymatic hydrolysis of wheat straw on cell wall composition, hydrophobicity and cellulase adsorption. *Bioresour. Technol.* **2011**, *102*, 5938–5946. [CrossRef] [PubMed]

40. Binks, B.; Lumsdon, S. Pickering emulsions stabilized by monodisperse latex particles: Effects of particle size. *Langmuir* **2001**, *17*, 4540–4547. [CrossRef]

41. Pawar, A.B.; Caggioni, M.; Ergun, R.; Hartel, R.W.; Spicer, P.T. Arrested coalescence in Pickering emulsions. *Soft Matter* **2011**, *7*, 7710–7716. [CrossRef]

42. Binks, B.P.; Fletcher, P.D.I.; Holt, B.L.; Beaussoubre, P.; Wong, K. Phase inversion of particle-stabilised perfume oil-water emulsions: Experiment and theory. *Phys. Chem. Chem. Phys.* **2010**, *12*, 11954–11966. [CrossRef] [PubMed]

43. Binks, B.P.; Lumsdon, S.O. Effects of oil type and aqueous phase composition on oil-water mixtures containing particles of intermediate hydrophobicity. *Phys. Chem. Chem. Phys.* **2000**, *2*, 2959–2967. [CrossRef]

materials

MDPI

Article

Quaternized Cellulose Hydrogels as Sorbent Materials and Pickering Emulsion Stabilizing Agents

Inimfon A. Udoetok [1], Lee D. Wilson [1,*] and John V. Headley [2]

[1] Department of Chemistry, University of Saskatchewan, 110 Science Place, Saskatoon, SK S7N 5C9, Canada; Inimfon.udoetok@usask.ca

[2] Water Science and Technology Directorate, Environment and Climate Change Canada, 11 Innovation Boulevard, Saskatoon, SK S7N 3H5, Canada; john.headley@canada.ca

* Correspondence: lee.wilson@usask.ca; Tel.: +1-306-966-2961; Fax: +1-306-966-4730

Academic Editor: To Ngai
Received: 30 June 2016; Accepted: 22 July 2016; Published: 30 July 2016

Abstract: Quaternized (QC) and cross-linked/quaternized (CQC) cellulose hydrogels were prepared by cross-linking native cellulose with epichlorohydrin (ECH), with subsequent grafting of glycidyl trimethyl ammonium chloride (GTMAC). Materials characterization via carbon, hydrogen and nitrogen (CHN) analysis, thermogravimetric analysis (TGA), and Fourier transform infrared (FTIR)/^{13}C solid state NMR spectroscopy provided supportive evidence of the hydrogel synthesis. Enhanced thermal stability of the hydrogels was observed relative to native cellulose. Colloidal stability of octanol and water mixtures revealed that QC induces greater stabilization over CQC, as evidenced by the formation of a hexane–water Pickering emulsion system. Equilibrium sorption studies with naphthenates from oil sands process water (OSPW) and 2-naphthoxy acetic acid (NAA) in aqueous solution revealed that CQC possess higher affinity relative to QC with the naphthenates. According to the Langmuir isotherm model, the sorption capacity of CQC for OSPW naphthenates was 33.0 mg/g and NAA was 69.5 mg/g. CQC displays similar affinity for the various OSPW naphthenate component species in aqueous solution. Kinetic uptake of NAA at variable temperature, pH and adsorbent dosage showed that increased temperature favoured the uptake process at 303 K, where Q_m = 76.7 mg/g. Solution conditions at pH 3 or 9 had a minor effect on the sorption process, while equilibrium was achieved in a shorter time at lower dosage (ca. three-fold lower) of hydrogel (100 mg vs. 30 mg). The estimated activation parameters are based on temperature dependent rate constants, k_1, which reveal contributions from enthalpy-driven electrostatic interactions. The kinetic results indicate an ion-based associative sorption mechanism. This study contributes to a greater understanding of the adsorption and physicochemical properties of cellulose-based hydrogels.

Keywords: cellulose; hydrogel; sorption; cross-linking; quaternization; hydrophile–lipophile balance; cooperative interactions

1. Introduction

Hydrogels are three-dimensional macromolecular polymer networks capable of absorbing and retaining significant water content due to its hydrophilic nature and network structure [1–3]. Hydrogels are usually prepared by various methods such as chemical cross-linking [4] and physical entanglement [5]. Biopolymer hydrogels derived from polysaccharides are very promising as carrier systems for drug delivery, stabilizers for Pickering emulsions, and adsorbents for pollutant removal. Biopolymer hydrogels are advantageous due to their abundance, low toxicity, biocompatibility, biodegradability, and tunable functionality [3,6–8]. Cellulose is an abundant and renewable biomaterial, white coloration, odourless, and nontoxic in nature. Cellulose also has unique properties such as mechanical strength, biocompatibility, hydrophilicity, relative thermal stability and tunable functionality [6]. The superior mechanical strength of cellulose over other polysaccharides with similar chemical structure and functionality relate to the strong versatile hydrogen bonding within and between biopolymer strands. This structural arrangement allows for segregation of hydrophobic and hydrophilic regions of the polymer to interact favourably with other components [9–11].

Cross-linking of cellulose-based hydrogels was reported using insoluble cellulose derivatives with bifunctional linkers such as epichlorohydrin [12], epichlorohydrin with aqueous ammonia [13], and divinylsulfone (DVS) [14]. As well, water soluble forms of cellulose containing quaternary ammonium salt derivatives [15,16] have also been explored. Previous reports [12,17] indicate that the point of zero charge (PZC) of cellulose-based hydrogels range from 6 to 8; thus, affording suitable uptake properties of both anionic and cationic adsorbates according to the pH conditions. However, these hydrogels are used for the sorption of anions such as naphthenic acid fraction components (NAFCs) found in oil sands process water (OSPW). At pH conditions above the zeta potential of the hydrogel, adsorptive uptake is limited due to electrostatic repulsion between the negatively charged hydrogel surface and the ionized naphthenates. However, the presence of hydrophilic moieties on the hydrogel surface may enhance the uptake affinity with amphiphilic naphthenate species, according to the "law of matching water affinities" [18]. OSPW is reported to display toxicity [19–24] due to the water soluble organic fraction that contains the major persistent toxic constituent, referred to as the NAFCs. NAFCs exhibit chemical stability, surfactant-like properties, persistence in the environment, non-volatility and high viscosity, especially at pH values above the pK_a of NAFCs. The growth in the oil sands extraction industry and the zero-discharge policy of the Government of Canada indicate the need for the development of low cost and sustainable technology for the treatment of the process-affected water. Biopolymer sorbents and their modified hydrogel forms may serve to address this need by subsequent reclamation of the vast area of land used for the storage of these tailings and process-affected water [25,26].

One approach to improve the affinity of cellulose hydrogels for contaminants like NAFCs at pH values above their point of zero charge (PZC) is to cross-link with bifunctional units such as epichlorohydrin before further grafting with quaternary ammonium salts such as glycidyl trimethyl ammonium chloride (GTMAC). Cross-linking with epichlorohydrin may enhance the surface area, porosity and hydrophobic character of the hydrogel [12], while grafting with quaternary ammonium salts will alter the hydrophile–lipophile balance (HLB) of the system. The shift of the PZC to higher pH values through the introduction of the quaternary ammonium groups affords charge stabilization of the hydrogel over a wider range of pH conditions [27].

The aim of this research was focused on understanding the role of cross-linking of cellulose with epichlorohydrin and surface modification via grafting with NH_4OH and GTMAC, along with a study of the uptake properties of hydrogel materials with NAFCs. This research will contribute to the development of low cost and sustainable materials for the treatment of tailing ponds in several ways: (*i*) to provide a greater understanding of the chemistry of cellulose hydrogels by studying the effect of cooperative interactions for hydrogel/adsorbate systems; (*ii*) to study the effect of cellulose cross-linking before the introduction of quaternary ammonium ions on the adsorption properties of such hydrogel systems; and (*iii*) to evaluate the utility of cellulose hydrogels as Pickering emulsion stabilizers and as sorbent materials for the uptake of NAFCs.

2. Results and Discussion

2.1. Characterization of Cellulose Hydrogels

2.1.1. FTIR Studies

The Fourier transform infrared (FTIR) spectra of the cellulose hydrogels and native cellulose are displayed in Figure 1a. The salient features of the spectra include a broad band (~3000–3600 cm^{-1}) attributed to intermolecular bonded OH and NH groups, C–H stretching (~2800–3000 cm^{-1}), O–H and C–H bending (~1400–1300 cm^{-1}), and C–O–H and C–O–C asymmetric stretching (~1000–1200 cm^{-1}), in agreement with a study on cross-linked cellulose [12]. The spectra reveal differences between native cellulose and the hydrogels, where the introduction of methylene, methyl and quaternary ammonium groups occur due to cross-linking and/or quaternization reactions, in agreement with reported studies [28,29]. The different features include IR bands at 1485 cm^{-1}, ~1000–1200 cm^{-1} and ~2934 cm^{-1} which display greater intensity relative to native cellulose, along with attenuation of the bands at ~1300–1400 cm^{-1}. The new IR bands relate to the methyl groups from the quaternary ammonium ion, ether linkage from cross-linking with epichlorohydrin, as well as the methylene groups due to cross-linking and/or quaternization reactions, in agreement with other reports [30,31]. An increased intensity and broadening of the band at ~1640 cm^{-1} relates to adsorbed water molecules and the C–N band at ~1450 cm^{-1} for the hydrogels provide further evidence for the introduction of quaternary ammonium groups onto the surface sites of cellulose [17,30].

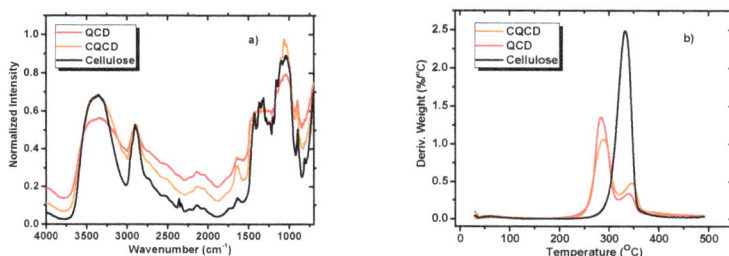

Figure 1. (**a**) FTIR spectra; and (**b**) DTG plot of cellulose and the hydrogels.

2.1.2. Thermogravimetry Studies

The thermal stability of native cellulose and the hydrogels are inferred from the DTG plot of derivative weight (%/°C) vs. temperature (°C) in Figure 1b. The thermogravimetric analysis (TGA) results show a single thermal event for cellulose relative to two thermal events for the hydrogels, in agreement with an independent report [32]. The maximum decomposition temperature for pristine cellulose occurs at ~320 °C; whereas, the hydrogels display maximum decomposition at ~260 °C and ~360 °C. The two thermal events for the hydrogels differ compared to cellulose, where a single thermal event is shown. The TGA results of the hydrogels relate to cross-linking and/or quaternization of cellulose, in agreement with the FTIR band at 1485 cm^{-1} in Figure 1a. The lower temperature event (~260 °C) for the hydrogels correspond to the cross-linker and quaternary ammonium groups; whereas, the event at ~360 °C relates to the decomposition of cellulose [32]. The DTG profiles also reveal that the intensity of the thermal event due to the decomposition of the cross-linker and quaternary ammonium groups of the hydrogel differ from the profile of cellulose. The foregoing provides support that cellulose is modified via cross-linking and quaternization. The profile also reveals that the cross-linked and quaternized hydrogel (CQC) has greater thermal stability over the quaternized hydrogel (QC). This greater stability of the cross-linked and quaternized hydrogel may relate to cross-linking effects, as noted for cellulose in another study [12].

2.1.3. Carbon, Hydrogen and Nitrogen (CHN) Composition of Hydrogels

The CHN results of the hydrogel are compared with cellulose in Table 1. The hydrogel contains greater C, H, and N content than unmodified cellulose. The presence of nitrogen in the hydrogels confirms the grafting of quaternary ammonium groups, in agreement with the above FTIR and DTG results, along with a related study [33]. The results reveal that CQC contains more nitrogen than QC and may be a consequence of the cross-linking of cellulose with ECH and grafting with aqueous ammonia before the quaternization process. The effect corresponds with a related study for the cross-linking of hydroxypropyl cellulose with ECH and aqueous ammonia [13].

Table 1. CHN content (%) and solvent swelling results of cellulose and its modified forms.

Material	% C	% H	% N	Swelling (Water)%	Swelling (Octanol)%	HLB *
Cellulose	41.0	6.27	NA	125	NA	NA
CQC	42.4	6.48	1.17	292	211	0.720
QC	42.8	7.43	0.80	415	260	0.625

* HLB is estimated by proxy as the ratio of swelling in octanol: water.

2.1.4. Hydrophile–Lipophile Balance (HLB) of Cellulose Hydrogels

The relative HLB of the cellulose materials was estimated by proxy, according to results obtained for solvent swelling in octanol and water (cf. Table 1 and Figure 2). The swelling results reveal that QC without cross-linking displayed greater swelling in water and octanol relative to its cross-linked form (CQC). The ratio of the swelling in octanol and water reveal that QC has measureable differences in HLB over CQC (CQC = 0.720; QC = 0.625). The HLB results are affirmed by QC which displays the greatest stabilization of a hexane in water Pickering emulsion system (cf. Figure 2a), exceeding four weeks. The colloidal stabilization of the emulsion by QC may be due to its greater wetting properties and ability to reduce the interfacial energy by adsorption at the oil/water interface [34,35]. The foregoing also reveals that CQC is a more hydrophobic hydrogel as evidenced by its HLB value, in agreement with the greater sorptive uptake of 2-naphthoxy acetic acid (NAA) (cf. Figure 2b), an amphiphilic adsorbate.

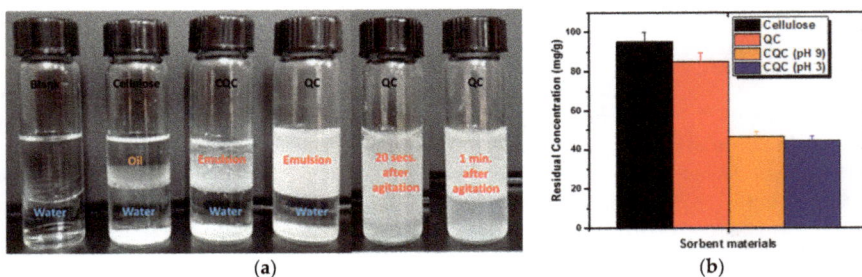

Figure 2. (**a**) Stabilization of oil/water emulsions by cellulose and hydrogels; and (**b**) uptake of NAA by cellulose and the hydrogels at pH 3 and 9 at 293 K.

2.1.5. ^{13}C NMR Studies of Cellulose Hydrogels

The ^{13}C solids NMR spectra of native cellulose and the hydrogels are presented in Figure 3. Unmodified cellulose display the following ^{13}C resonances: C1 (105 ppm), C2/C3/C5 (68–78 ppm), C4 (88.4 and 83.3 ppm), and C6 (57–67 ppm), in agreement with a previous report [36]. The ^{13}C spectra of the hydrogels possess features that resemble native cellulose, in agreement with the component nature of the cellulose biopolymer structural units. The hydrogels reveal new ^{13}C NMR resonance lines, band broadening, and chemical shift variations. These effects are absent in the spectra of cellulose, in accordance with the DTG and FTIR results above. For example, new resonance lines at ~55.0 ppm for $(CH_3)_3N^+$ and ~70–77 ppm relate to CH_2 groups from ECH and GTMAC, respectively, in agreement with other reports [37,38]. The resonance lines for ECH and GTMAC may overlap with the spectral features (cf. ~62–65 ppm and ~68–78 ppm) of native cellulose, in accordance with the line broadening related to quaternization [38]. The shoulder at ~103 ppm may be due to C3 which bears substituted or non-substituted hydroxyl groups [39]. The difference between CQC and QC in the ^{13}C NMR spectra relate to the well resolved and increased spectral intensity of the lines ~68–78 ppm from ECH and GTMAC, related to their similar chemical environments. QC was synthesized via grafting cellulose with GTMAC without cross-linking. The ^{13}C NMR results provide support that cross-linking occurs between cellulose and ECH with quaternization due to GTMAC, in agreement with the above DTG results.

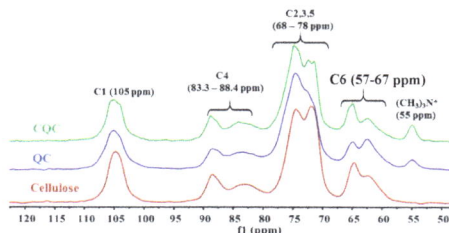

Figure 3. ^{13}C CP-MAS solids NMR spectra of cellulose, QC and CQC obtained at 125.8 MHz with MAS at 10 kHz and 293 K.

2.2. Sorption Studies

2.2.1. Comparative Uptake of Single Component Carboxylate Anion

Sorption results for the uptake affinity of cellulose and the hydrogels with NAA in aqueous solution are shown in Figure 2b. The results reveal that CQC displays greater affinity for the uptake of NAA relative to cellulose and QC. The greater affinity of CQC with NAA likely relate to the *surfactant-like* properties of NAA, in accordance with the HLB results presented above for CQC. Cross-linking of cellulose with ECH results in reduced water solubility; whereas, grafting with GTMAC further enhances the hydrophile character due to the introduction of quaternary ammonium groups. Thus, the uptake of NAA by CQC is a result of cooperative interactions. The uptake results in Figure 2b shows that pH has a minor effect on the equilibrium uptake of NAA by CQC due to the grafting of the quaternary ammonium ions onto cellulose. Previous reports on the uptake properties [12] of anions with cellulose materials indicate that low anion affinity occurs at alkaline pH as a result of the negative zeta-potential on the sorbent surface [40]. The introduction of quaternary ammonium groups offset the low anion affinity in the case of unmodified cellulose through the introduction of a positive zeta-potential, as for the case of QC or CQC.

2.2.2. Equilibrium Studies of Single Component Carboxylate Anion (NAA) and OSPW Naphthenates

The uptake isotherms of the sorbents with NAA and OSPW naphthenates are presented as plots of Q_e versus C_e in Figure 4a. The isotherms display a non-linear increase for Q_e as C_e increases. The Langmuir model provides a reasonable "best-fit" to the experimental results and reveal that the sorbent surface has homogenous sorption sites (cf. Table 2). The proposed structure of CQC in Scheme 1 agrees with the ^{13}C solids NMR and FTIR spectral results, where two possible sorption sites exist in such sorbent materials. These include the quaternary ammonium sites and the hydrophobic domains of cellulose even though the isotherm results are described by the Langmuir model. Evidence of a singular type of sorption site for these materials can be understood on the basis of a large offset in the binding affinity of each respective site. The uptake of NAA and OSPW naphthenates by CQC is likely dominated by the ion–ion electrostatic attractions, aided by cooperative hydrophobic effects. The Langmuir isotherm data shows that CQC displayed a two-fold greater uptake of NAA (Q_m = 69.5 mg/g) over OSPW naphthenates (Q_m = 33.0 mg/g). The variable uptake of NAA by CQC may be due to the mixed composition of OSPW, where certain congeners have reduced uptake due to steric effects or variable HLB that limit the binding with CQC. The variable lipophilicity profile of OSPW relates to the z-value of the naphthenates, where the z-value denotes the "hydrogen deficiency", as reported elsewhere [41–45]. Naphthenates with z-values less than 2 in the OSPW display enhanced binding affinity with CQC, as follows: OSPW (0.0333 L/mg) and NAA (0.234 L/mg). The hydrogel reported herein has favourable uptake relative to activated carbon (AC) and nickel-based alumina (Ni-Al$_2$O$_3$), 20 mg/g, and chemically treated AC, 35 mg/g, for OSPW naphthenates [46,47].

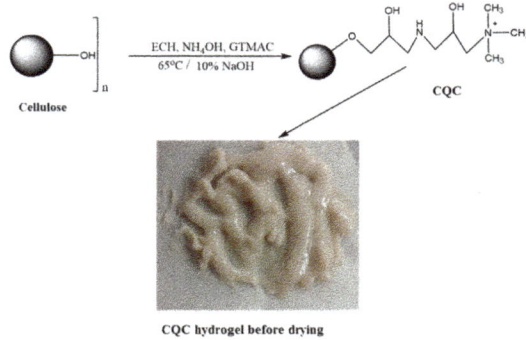

CQC hydrogel before drying

Scheme 1. Synthetic scheme for the preparation of cross-linked/quaternized CQC.

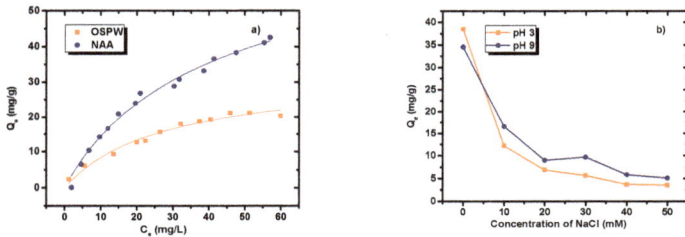

Figure 4. (**a**) Sorption isotherm of CQC with NAA at pH 9 and OSPW at pH 10.5 and 293 K; and (**b**) effect of ion concentration on the sorption isotherm of NAA by CQC at pH 3 and 9 and 293 K.

Table 2. Sorption isotherm parameters obtained from the Langmuir, Sips, and Freundlich models for CQC with OSPW naphthenates (pH 10.5) and NAA (pH 9.0) at 293 K.

Adsorbates	Isotherm Model	Parameters	Sorbent (CQC)
OSPW	Langmuir	Q_m (mg·g^{-1})	33.0
		K_L (L·mg^{-1})	0.0333
		SSE	0.921
	Sips	Q_m (mg·g^{-1})	36.9
		K_S (L·mg^{-1})	0.0370
		n_s	0.904
		SSE	2.73
	Freundlich	K_F (L·mg·g^{-1})	2.63
		$1/n_f$	1.89
		SSE	2.75×10^3
NAA	Langmuir	Q_m (mg·g^{-1})	69.5
		K_L (L·mg^{-1})	0.0260
		SSE	0.994
	Sips	Q_m (mg·g^{-1})	60.5
		K_S (L·mg^{-1})	0.0234
		n_s	1.12
		SSE	4.10
	Freundlich	K_F (L·mg·g^{-1})	3.47
		$1/n_f$	2.09
		SSE	1.13×10^6

2.2.3. Effects of Ion Concentration on the Uptake of NAA

The concentration effect of NaCl on the sorption of NAA with CQC is presented in Figure 4b. The results reveal that an increase in the level of NaCl in the NAA solution results in reduced uptake of NAA with CQC, in agreement with other reports [46,48]. The uptake of NAA decreased from 38 mg/g to 5 mg/g at pH 3, and 35 mg/g to 6 mg/g at pH 9 as the level of NaCl increased to 50 mM. The attenuated uptake of NAA in the presence of electrolyte may be due to charge screening effects at the sorption sites between the carboxylate anions and the quaternary ammonium groups, in agreement with enhanced ion–ion binding contributions, as outlined above.

2.2.4. Sorption of OSPW Naphthenates by Cellulose Hydrogels

The ESI-HRMS speciation profile of OSPW before and after sorption with CQC is shown in Figure 5a,b. Figure 5a displays the class distribution plots for OSPW and show that O_2H species are the most prominent; whereas, the O_2S species are of secondary abundance. The results demonstrate a significant attenuation of the concentration of the O_2H species after sorption with the CQC hydrogel. The foregoing shows that the CQC hydrogel has favourable uptake affinity and removal of these OSPW fractions at equilibrium.

Figure 5b presents a distribution profile of double bond equivalent (DBE) species in OSPW as a function of the normalized concentration of O_2 species. The DBE accounts for the hydrogen deficiency due to ring formation or double bonds of the naphthenate systems [49]. The results show that the O_2H species are comprised of naphthenates with DBE values in the range of 1 to 10. In addition, the more prominent species in the OSPW mixture have DBE values between 2 and 8. The uptake results show that CQC has little or no selectivity toward the carboxylate anion species in the mixture, in agreement with the electrostatic driven binding for such hydrogel/OSPW systems. The removal efficiency ranged from 38% (DBE = 4) to 69% (DBE = 8), while Mohamed et al. [50] reported an increase in removal efficiency with increasing DBE due to hydrophobic effects. Udoetok et al. [12] reported greater uptake of model naphthenates as the number of rings increased for cellulose and its cross-linked forms. The foregoing shows that CQC has appreciable fractionation efficacy for OSPW naphthenates that may relate to cooperative electrostatic and hydrophobic effects, as noted above.

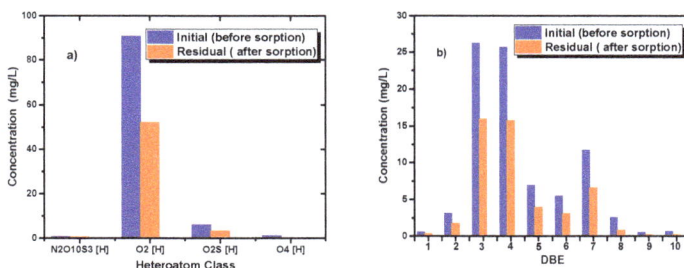

Figure 5. (**a**) Electrospray ionization high resolution mass spectrometry (ESI-HRMS) speciation profile of OSPW; and (**b**) double bond equivalents (DBE) distribution of OSPW as a function of normalized concentration for the O_2 species before and after sorption with CQC at pH 10.5 and 293 K.

2.3. *Kinetic Uptake Studies*

2.3.1. Effects of Temperature and Sorbent Dosage

Figure 6 illustrates the effects of temperature on the kinetic uptake of NAA from aqueous solution by CQC via the one-pot kinetic system. The uptake profile is well described by the pseudo-first order (PFO) kinetic model, in agreement with a single type of sorption site. The results show an increase in Q_t values with increasing time, where saturation occurs at t > 400 s. In Figure 6a, the kinetic profile reveals that Q_e (mg/g) and the PFO rate constant, k_1 (s^{-1}) increase with temperature at a low dosage (30 mg) of CQC (cf. Table 3), where dynamic equilibrium is reached after 460 min. At higher hydrogel dosage (100 mg) in Figure 6b, the rate becomes attenuated as temperature increases, due to the fact that the monolayer saturation of the biopolymer was not met after 400 min. A comparison of the Q_e values for each adsorbent dosage (cf. Table 4) reveal that uptake of CQC (30 mg dosage) at 303 K was higher relative to uptake for the hydrogel (100 mg dosage), in support of the claim that monolayer saturation was not achieved after 400 s. Figure 6b shows that the uptake profile at 303 K converged with the results at 298 K, supporting the claim that longer time is required to achieve dynamic equilibrium. The increase of Q_e and k_1 values as temperature increases indicate that higher temperature provides the required energy to activate the adsorption process, in agreement with other results [51].

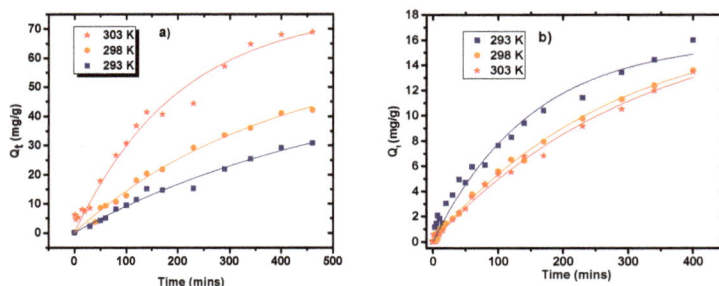

Figure 6. Kinetic uptake profile of NAA: (**a**) low dosage (30 mg); and (**b**) high dosage (100 mg) of CQC at pH 9 and 293 K, 298 K and 303 K. The fitted lines correspond to the PFO model.

Table 3. Pseudo-first order (PFO) kinetic uptake results of NAA by CQC at 293 K, 298 K and 303 K and pH 3 and 9.

Weight of Adsorbent	Temperature (K)	pH	Parameters		
			Q_m (mg/g)	K_1 (S^{-1})	R^2
30	293	9	54.2	0.00184	0.980
	298		61.3	0.00265	0.992
	303		76.7	0.00491	0.982
100	293		16.0	0.00666	0.974
	298		17.4	0.00634	0.997
	303		18.0	0.00319	0.992
	293	3	23.2	0.00943	0.996

Table 4. Thermodynamic parameters for the uptake of NAA by CQCD.

Temp (K)	ΔE_a (kJ/mol)	Activation Parameters		
		ΔH^* (kJ/mol)	ΔS^* (J/Kmol)	ΔG^* (kJ/mol)
293				57.9
298	72.3	0.291	−196	58.9
303				59.8

2.3.2. Effects of pH

A study of pH effects on the uptake kinetics of NAA by CQC hydrogel is shown in Figure 7. The results show an increase for Q_t as time increases that agree with general effects of temperature and sorbent dosage. The kinetic uptake results at pH 3 (Q_e) and the PFO rate constant (k_1) are greater relative to conditions at pH 9 (cf. Table 3). These results do not agree with the comparative equilibrium uptake of NAA by cellulose and the hydrogels at pH 3 and 9. The slight disparity in the uptake capacity between the equilibrium and kinetic studies is related to the earlier claim that 400 min is insufficient for achieving isotherm saturation, and the faster rate of the uptake at pH 3. The greater PFO rate constant obtained for the kinetic uptake at pH 3 is related to the reduced role of counter ion binding with the quaternary ammonium ions at acidic pH relative to pH 9, where greater charge screening effects are anticipated as the level of OH⁻ increases.

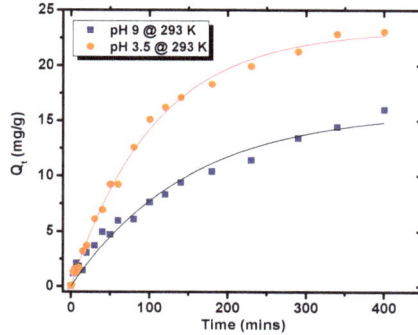

Figure 7. Kinetic uptake profile of NAA at 293 K with a 0.833 mg/mL dosage of CQC at pH 9 and 3, where the fitted lines correspond to the PFO model.

2.3.3. Activation Parameters

The calculated activation parameters such as the change in Gibbs energy (ΔG), enthalpy change (ΔH), entropy change (ΔS), and activation energy (ΔE_a), were based on the trend in PSO rate constants (k_1) with temperature. The activation parameters were calculated from a plot of ln k_1/T versus $1/T$ (cf. Figure 8b) by the Eyring equation (Equation (9)). The positive value of ΔG^* (cf. Table 4) at variable temperature shows a decrease with increasing temperature, indicating that the sorption process is exergonic. The endothermic nature of ΔH^* of the sorption process is anticipated for electrostatic driven processes. Based on the ΔS^* values, an associative sorption mechanism is inferred which proceeds via the formation of an activated complex for the hydrogel system [52]. The ΔE_a value in Table 4 derived from the plot of ln k_1 versus $1/T$ (cf. Figure 8a) by the Arrhenius equation (Equation (11)) is 72 kJ/mol. The result agrees with estimates of E_a for other ion exchange systems reported elsewhere [53]. The foregoing along with the Langmuir-based description for the adsorption process support the claim that electrostatic interactions are predominant with evidence of cooperative hydrophobic effects for the uptake of NAA by CQC.

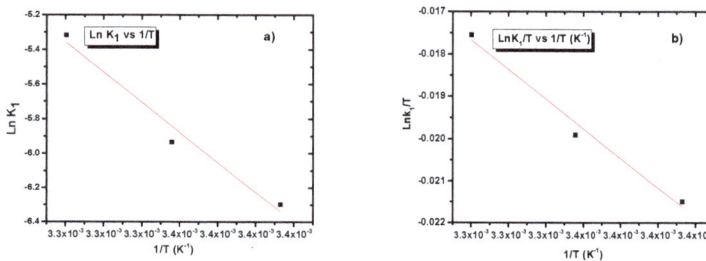

Figure 8. (**a**) Plot of ln k_1 versus $1/T$ for the determination of activation energy (ΔE_a); and (**b**) plot of ln k_1/T versus $1/T$ for the determination of activation parameters of adsorption for the CQC–NAA system.

3. Materials and Methods

3.1. Materials

Cellulose (medium fibre from cotton linters), sodium hydroxide, aqueous ammonia, sodium chloride, 99% epichlorohydrin (ECH) glycidyl trimethyl ammonium chloride (GTMAC), HCl, and ACS grade acetone and octanol were obtained from Sigma Aldrich (St. Louis, MO, USA). All materials were used as received without further purification.

3.2. Synthesis of Cellulose Hydrogels

Synthesis of a cellulose hydrogel (CQC) was carried out using 2 g of bulk cellulose with heating in 15 mL of 10% NaOH at 65 °C for 3 h. One millilitre of ECH followed by 1 mL of NH_4OH was added to the NaOH-cellulose mixture in a drop-wise manner over a one-minute period, where the reaction mixture was stirred at 65 °C for ca. 16 h. The resulting reaction mixture was neutralized using 6 M HCl and the product separated from the supernatant via vacuum filtration, followed by washing with cold Millipore water and ACS grade acetone. The resulting product was added to another round bottom flask and heated in 15 mL of 10% NaOH 65 °C for 1 h followed by the drop-wise addition of 10 mL glycidyl trimethyl ammonium chloride (GTMAC). The reaction mixture was allowed to stir for at least 6 h before separation of the final products via vacuum filtration, followed by washing with cold Millipore water and ACS grade acetone with drying at 60 °C. Soxhlet extraction for 24 h with HPLC grade acetone was used to remove unreacted/excess reagents, followed by drying in a vacuum oven at 60 °C for 12 h. The resulting polymer was ground in a mortar and pestle and sieved with a 40-mesh sieve. Another hydrogel (QC) was also synthesized via the same method but without the ECH and NH_4OH for the cross-linking step.

3.3. Characterization

3.3.1. Fourier Transform Infrared (FTIR) Spectroscopy

The FTIR spectra of the hydrogels and cellulose were obtained using A Bio-RAD FTS-40 IR spectrophotometer (Bio-Rad Laboratories, Inc., Philadelphia, PA, USA). Dry and powdered samples were mixed with pure spectroscopic grade KBr in a weight ratio of 1:10 with grinding in a small mortar and pestle. The DRIFT (Diffuse Reflectance Infrared Fourier Transform) spectra were obtained in reflectance mode at room temperature with a resolution of 4 cm^{-1} over the 400–4000 cm^{-1} spectral range. Multiple scans were recorded and corrected relative to a background of pure KBr.

3.3.2. Thermal Gravimetric Analysis (TGA)

Thermal stability of the hydrogels and cellulose were determined using a TA Instruments Q50 TGA system (New Castle, DE, USA) operated with a heating rate of 5 °C min^{-1} up to 500 °C using nitrogen as the carrier gas. The results obtained are shown as first derivative (DTG) plots of weight with temperature (%/°C) against temperature (°C).

3.3.3. Carbon, Hydrogen and Nitrogen (CHN) Analyses

The C, H, and N composition of the hydrogels and cellulose were obtained using a Perkin Elmer 2400 CHN Elemental Analyzer (PerkinElmer, Inc., Waltham, MA, USA) with the following instrument conditions: combustion oven temperature (above 925 °C) and reduction oven temperature (above 640 °C). The instrument was purged with a mixture of pure oxygen and helium gas, where acetanilide was used as the calibration standard. Elemental analyses were obtained in triplicate with an estimated precision of ±0.3%.

3.3.4. Hydrophile–Lipophile Balance (HLB)

The hydrophile–lipophile balance (HLB) of the hydrogels was estimated via swelling in Millipore water and octanol, respectively. This was done by shaking approximately 30 mg of the materials in 15 mL of Millipore water and octanol in a horizontal shaker for ~48 h. The weights of swollen hydrogels (w_s) were determined after tamping dry with filter paper. The dry weights (w_d) were obtained after drying in an oven at 65 °C to a constant mass. The swelling ratio was calculated for each neat solvent system using Equation (1):

$$Sw\,(\%) = \frac{Ws - Wd}{Wd} \times 100 \tag{1}$$

3.3.5. Solid State ^{13}C NMR Spectroscopy

A Bruker AVANCE III HD spectrometer (Bruker Bio Spin Corp., Billerica, MA, USA) furnished with a 4 mm DOTY CP-MAS (cross polarization with magic angle spinning) solids probe operating at 125.8 MHz (^{1}H frequency at 500.2 MHz) was use to acquire the ^{13}C solids NMR spectra of the hydrogels and cellulose. The experimental conditions were as follows: spinning speed of 10 kHz, a ^{1}H 90° pulse of 3.5 μs, a contact time of 0.75 ms, with a ramp pulse on the ^{1}H channel. Others are MAS rate of 10 kHz, a ^{13}C 90° pulse of 3.15 μs and a 25 kHz SPINAL-64 ^{1}H decoupling during acquisition. For different samples, 600–5000 scans were accumulated, with a recycle delay of 2 s. All experiments were recorded using 71 kHz SPINAL-64 decoupling during acquisition. Chemical shifts were referenced to adamantane at 38.48 ppm (low field signal).

3.4. Sorption Studies

3.4.1. Sorption of OSPW Naphthenates and Single Component Carboxylate Ions

A 100-mL stock solution at 100 ppm was prepared for the OSPW and 150 ppm for 2-naphthoxy acetic acid (NAA), respectively. An appropriate amount of NAA was dissolved in an aqueous NH_3 solution with sonication and further stirring until the resulting solution was clear, while 2800 ppm OSPW was diluted using Millipore water and stirred overnight. Solutions with variable concentration (1–150 ppm) were prepared by dilution of the stock with Millipore water.

Fixed amounts (10 mg) of the hydrogels were mixed with 5 mL of NAA and OSPW solutions in 2 dram vials at variable concentration. The mixtures were equilibrated at room temperature on a horizontal shaker table for 24 h. The initial concentration (C_o) and residual concentration at equilibrium (C_e) of OSPW naphthenates and NAA were determined using an electrospray ionization high resolution mass spectrometer and a double beam spectrophotometer (CARY 100, Varian, Mulgrave, Australia) at room temperature (293 ± 0.5 K). The samples were centrifuged prior to ESI-HRMS and UV-vis spectroscopy analyses. Uptake of NAA and NAFCs in OSPW was determined from the difference between C_o and C_e values described by Equation (2).

$$Q_e = \frac{(C_o - C_e) \times V}{m} \tag{2}$$

Q_e is the quantity of adsorbate uptake in the solid phase at equilibrium (mg·g^{-1}); C_o is initial concentration of adsorbate (mg·L^{-1}) in solution; C_e is concentration of adsorbate at equilibrium (mg·L^{-1}) in solution; V is volume of adsorbate solution; and m is the mass (g) of sorbent.

3.4.2. Electrospray Ionization Mass Spectrometric (ESI-HRMS) Quantification

The residual (C_e) and initial (C_o) concentration of the NAFCs was estimated using a ThermoScientific LTQ Orbitrap Elite Electrospray ionization high resolution mass spectrometer (ESI-HRMS) (Thermo Fisher Scientific Inc., Waltham, MA, USA). The resolution setting of the spectrometer was 30,000 while a full-scan mass spectrum was collected between *m/z* 100 and 600.

Quantification of samples was achieved by extracting the mass range of the analyte of interest. The electrospray ionization (ESI) interface was set to negative-ion mode. Mass spectrometer conditions were optimized by the transmission of *m/z* 112.98563. The heated ESI interface (HESI) parameters were as follows: source heater temperature (53 °C); spray voltage (2.86 kV); capillary temperature (275 °C); sheath gas flow rate (25 $L \cdot h^{-1}$); auxiliary gas flow rate (5 $L \cdot h^{-1}$); and spray current (5.25 μA).

3.4.3. Sorption Isotherms and Modeling

Sorption isotherms were obtained by plotting Q_e vs. C_e (cf. Equation (2)). The isotherms were subsequently analyzed using Langmuir [54] Freundlich [55] and the Sips [56] models (cf. Equation (3)–(5)). The "best fit" for the data was obtained by minimizing the SSE (cf. Equation (6)) for all data across the range of conditions. Q_{ei} is the experimental value, Q_{ef} is the calculated value from data fitting and N is the number of Q_e data points.

$$Q_e = \frac{K_L Q_m C_e}{1 + K_L C_e} \tag{3}$$

$$Q_e = K_F C_e^{1/n_f} \tag{4}$$

$$Q_e = \frac{Q_m (K_S C_e)^{n_s}}{1 + (K_S C_e)^{n_s}} \tag{5}$$

$$SSE = \sqrt{\frac{(Q_{ei} - Q_{ef})^2}{N}} \tag{6}$$

3.5. Kinetic and Thermodynamic Studies

3.5.1. Kinetic Studies

Kinetic studies were carried out using a one pot method [57] as follows: variable amounts (~100 mg or 30 mg) of the hydrogel were added to a folded Whatman filter paper (55 mm diameter), where both ends were clamped after encasing the polymer. The clamped filter paper containing the sample was immersed in a fixed volume (120 mL) of a 150-ppm NAA solution. Three millilitres of the NAA solution were sampled after designated time intervals. The residual concentration of the aliquots was determined using a double beam spectrophotometer (Varian CARY 100) at 293 ± 0.5 K. Uptake of NAA at each sampling time interval (t) was estimated according to Equation (7), where C_o and C_t refer to the surrogate concentration at t = 0 and variable t.

$$Q_t = \frac{(C_o - C_t) \times V}{m} \tag{7}$$

Kinetic uptake isotherms were obtained by plotting Q_t vs. time according to the pseudo-first order (PFO kinetic model (cf. Equation (8)).

$$Q_t = Q_e(1 - e^{-k_1 t}) \tag{8}$$

3.5.2. Thermodynamic Studies

The standard enthalpy of activation (ΔH*), entropy of activation (ΔS*), and the Gibbs energy of activation (ΔG*) in the adsorption process was calculated from a plot of ln k_1/T versus 1/T according to the Eyring equation (Equation (9)),

$$\frac{\ln k}{T} = \ln \frac{k_B}{h} + \frac{\Delta S^*}{R} - \frac{\Delta H^*}{RT} \tag{9}$$

where k_1 is the adsorption rate constant; k_B is the Boltzmann constant (1.3807×10^{-23} J·K^{-1}); h is Planck's constant (6.6261×10^{-34} J·s); R is the ideal gas constant (8.314 J·mol^{-1}·K^{-1}); and T is the temperature (K). The values of ΔH^* and ΔS^* were determined from the slope and intercept of a plot of ln k_1/T versus 1/T. The values obtained were used to compute ΔG^* from Equation (10) below:

$$\Delta G^* = \Delta H^* - T\Delta S^* \tag{10}$$

The activation energy (E_a) of the process was obtained by plotting *ln* k_1 versus 1/T according to the Arrhenius Equation (11),

$$\ln k_1 = \ln A + \frac{E_a}{R}\left(\frac{1}{T}\right) \tag{11}$$

where k_1 is the rate constant; A is the pre-exponential factor; E_a is the activation energy; R is the gas constant; and T is the temperature (K).

4. Conclusions

Quaternized cellulose hydrogels as cross-linked (CQC) and non-cross-linked forms (QC) were prepared. The grafting of GTMAC with cellulose results in the presence of quaternary ammonium groups, according to structural support via FTIR, CHN analysis, ^{13}C NMR and TGA. The hydrogels possess enhanced network structure and variable hydrophobic character due to cross-linking and grafting effect. The TGA results reveal enhanced thermal stability of the hydrogels relative to cellulose, while solvent swelling studies in octanol and water indicate that QC has variable hydrophile–lipophile balance (HLB) relative to CQC. QC in hexane–water mixtures afford the formation of stable oil/water Pickering emulsions. The sorption of OSPW naphthenates and NAA by the hydrogels revealed that CQC has greater affinity over QC, where the uptake is favoured by cooperative hydrophobic and electrostatic interactions. CQC displays similar affinity for the component species of OSPW naphthenates, where the PFO model provides best fit for the experimental kinetic uptake results. The kinetic uptake process is favoured by increased temperature while pH has a minor effect. The activation parameters were derived from temperature dependence of the rate constants, k_1, where an entropy driven sorption process appears to follow an associative ion-ion driven sorption mechanism. This study contributes to a greater understanding of the structure-function properties of cellulose hydrogels and their uptake properties with carboxylate anions. This study will catalyze further development of low-cost and versatile biopolymer materials for tunable Pickering emulsions, chemical fractionation, and diverse adsorption-based processes.

Acknowledgments: The authors are grateful to the Government of Saskatchewan (Ministry of Agriculture) through the Agriculture Development Fund (PROJECT#: 20110162) for supporting this research. The support and cooperation of the University of Saskatchewan and Environment and Climate Change Canada are gratefully acknowledged. The technical expertise and assistance of Jonathan Bailey and Kerry Peru are credited for the mass spectrometry studies.

Author Contributions: Lee D. Wilson and Inimfon A. Udoetok conceived and designed the experiments; Inimfon A. Udoetok performed the experiments and analyzed the data; Lee D. Wilson secured funding and John V. Headley contributed analysis tools; Inimfon A. Udoetok wrote the first draft of the paper with extensive editing by Lee D. Wilson, where John V. Headley provided final proofreading of the final manuscript.

Disclaimer: As there is a lack of representative analytical standard methodologies related to the analysis of total NAs or NAFCs, the results presented in this manuscript should be considered internally consistent, but may not be directly comparable to other results. For further information on the methods and analytical procedures used in this study, please contact the corresponding author.

Conflicts of Interest: The authors declare no conflict of interest.

References

1. Chang, C.; Zhang, L. Cellulose-based hydrogels: Present status and application prospects. *Carbohyd. Polym.* **2011**, *84*, 40–53. [CrossRef]

2. Chang, C.; Zhang, L.; Zhou, J.; Zhang, L.; Kennedy, J.F. Structure and properties of hydrogels prepared from cellulose in naoh/urea aqueous solutions. *Carbohyd. Polym.* **2010**, *82*, 122–127. [CrossRef]

3. Sannino, A.; Demitri, C.; Madaghiele, M. Biodegradable cellulose-based hydrogels: Design and applications. *Materials* **2009**, *2*, 353–373. [CrossRef]

4. Molina, M.J.; Gómez-Antón, M.R.; Piérola, I.F. Determination of the parameters controlling swelling of chemically cross-linked ph-sensitive poly(n-vinylimidazole) hydrogels. *J. Phys. Chem. B* **2007**, *111*, 12066–12074. [CrossRef] [PubMed]

5. Wong, J.E.; Díez-Pascual, A.M.; Richtering, W. Layer-by-layer assembly of polyelectrolyte multilayers on thermoresponsive p(nipam-co-maa) microgel: Effect of ionic strength and molecular weight. *Macromolecules* **2009**, *42*, 1229–1238. [CrossRef]

6. Qiu, X.; Hu, S. "Smart" materials based on cellulose: A review of the preparations, properties, and applications. *Materials* **2013**, *6*, 738–781. [CrossRef]

7. Cavalieri, F.; Chiessi, E.; Finelli, I.; Natali, F.; Paradossi, G.; Telling, M.F. Water, solute, and segmental dynamics in polysaccharide hydrogels. *Macromol. Biosci.* **2006**, *6*, 579–589. [CrossRef] [PubMed]

8. Zoppe, J.O.; Venditti, R.A.; Rojas, O.J. Pickering emulsions stabilized by cellulose nanocrystals grafted with thermo-responsive polymer brushes. *J. Colloid Interf. Sci.* **2012**, *369*, 202–209. [CrossRef] [PubMed]

9. Luo, X.; Zhang, L. New solvents and functional materials prepared from cellulose solutions in alkali/urea aqueous system. *Food Res. Int.* **2013**, *52*, 387–400. [CrossRef]

10. Bergenstråhle, M.; Wohlert, J.; Himmel, M.E.; Brady, J.W. Simulation studies of the insolubility of cellulose. *Carbohyd. Res.* **2010**, *345*, 2060–2066. [CrossRef] [PubMed]

11. Chaplin, M. Water Structure and Science: Cellulose. Available online: http://www1.lsbu.ac.uk/water/cellulose.html (accessed on 20 June 2016).

12. Udoetok, I.A.; Dimmick, R.M.; Wilson, L.D.; Headley, J.V. Adsorption properties of cross-linked cellulose-epichlorohydrin polymers in aqueous solution. *Carbohyd. Polym.* **2016**, *136*, 329–340. [CrossRef] [PubMed]

13. Yan, L.; Shuai, Q.; Gong, X.; Gu, Q.; Yu, H. Synthesis of microporous cationic hydrogel of hydroxypropyl cellulose (HPC) and its application on anionic dye removal. *CLEAN Soil Air Water* **2009**, *37*, 392–398. [CrossRef]

14. Marc, G.; Mele, G.; Palmisano, L.; Pulito, P.; Sannino, A. Environmentally sustainable production of cellulose-based superabsorbent hydrogels. *Green Chem.* **2006**, *8*, 439–444. [CrossRef]

15. Chen, J.-C.; Yeh, J.-T.; Chen, C.-C. Crosslinking of cotton cellulose in the presence of alkyl diallyl ammonium salts. I. Physical properties and agent distribution. *J. Appl. Polym. Sci.* **2003**, *90*, 1662–1669. [CrossRef]

16. Yang, S.P.; Fu, S.Y.; Li, X.Y.; Zhou, Y.M.; Zhan, H.Y. Preparation of salt-sensitive and antibacterial hydrogel based on quaternized cellulose. *Bioresources* **2010**, *5*, 1114–1125.

17. Hu, D.Y.; Wang, P.; Li, J.; Wang, L.J. Functionalization of microcrystalline cellulose with n,n-dimethyldodecylamine for the removal of congo red dye from an aqueous solution. *Bioresources* **2014**, *9*, 5951–5962. [CrossRef]

18. Vlachy, N.; Jagoda-Cwiklik, B.; Vácha, R.; Touraud, D.; Jungwirth, P.; Kunz, W. Hofmeister series and specific interactions of charged headgroups with aqueous ions. *Adv. Colloid Interface Sci.* **2009**, *146*, 42–47. [CrossRef] [PubMed]

19. Leishman, C.; Widdup, E.E.; Quesnel, D.M.; Chua, G.; Gieg, L.M.; Samuel, M.A.; Muench, D.G. The effect of oil sands process-affected water and naphthenic acids on the germination and development of arabidopsis. *Chemosphere* **2013**, *93*, 380–387. [CrossRef] [PubMed]

20. Lengger, S.K.; Scarlett, A.G.; West, C.E.; Rowland, S.J. Diamondoid diacids ('O4' species) in oil sands process-affected water. *Rapid Commun. Mass Spectrom.* **2013**, *27*, 2648–2654. [CrossRef] [PubMed]

21. Reinardy, H.C.; Scarlett, A.G.; Henry, T.B.; West, C.E.; Hewitt, L.M.; Frank, R.A.; Rowland, S.J. Aromatic naphthenic acids in oil sands process-affected water, resolved by gcxgc-ms, only weakly induce the gene for vitellogenin production in zebrafish (danio rerio) larvae. *Environ. Sci. Technol.* **2013**, *47*, 6614–6620. [CrossRef] [PubMed]

22. Yue, S.; Ramsay, B.A.; Wang, J.; Ramsay, J. Toxicity and composition profiles of solid phase extracts of oil sands process-affected water. *Sci. Total Environ.* **2015**, *538*, 573–582. [CrossRef] [PubMed]

23. Wiseman, S.B.; Anderson, J.C.; Liber, K.; Giesy, J.P. Endocrine disruption and oxidative stress in larvae of chironomus dilutus following short-term exposure to fresh or aged oil sands process-affected water. *Aquat. Toxicol.* **2013**, *142–143*, 414–421. [CrossRef] [PubMed]

24. He, Y.; Patterson, S.; Wang, N.; Hecker, M.; Martin, J.W.; El-Din, M.G.; Giesy, J.P.; Wiseman, S.B. Toxicity of untreated and ozone-treated oil sands process-affected water (ospw) to early life stages of the fathead minnow (pimephales promelas). *Water Res.* **2012**, *46*, 6359–6368. [CrossRef] [PubMed]

25. Wilson, L.D.; Mohamed, M.H.; Headley, J.V. Novel materials for environmental remediation of oil sands contaminants. *Rev. Environ. Health* **2014**, *29*, 5–8. [CrossRef] [PubMed]

26. Tang, J.; Quinlan, P.J.; Tam, K.C. Stimuli-responsive pickering emulsions: Recent advances and potential applications. *Soft Matter* **2015**, *11*, 3512–3529. [CrossRef] [PubMed]

27. Oyanedel-Craver, V.A.; Smith, J.A. Effect of quaternary ammonium cation loading and ph on heavy metal sorption to ca bentonite and two organobentonites. *J. Hazard Mater.* **2006**, *137*, 1102–1114. [CrossRef] [PubMed]

28. Song, Y.; Zhou, J.; Li, Q.; Guo, Y.; Zhang, L. Preparation and characterization of novel quaternized cellulose nanoparticles as protein carriers. *Macromol. Biosci.* **2009**, *9*, 857–863. [CrossRef] [PubMed]

29. Oh, S.Y.; Yoo, D.I.; Shin, Y.; Seo, G. Ftir analysis of cellulose treated with sodium hydroxide and carbon dioxide. *Carbohyd. Res.* **2005**, *340*, 417–428. [CrossRef] [PubMed]

30. Pei, A.; Butchosa, N.; Berglund, L.A.; Zhou, Q. Surface quaternized cellulose nanofibrils with high water absorbency and adsorption capacity for anionic dyes. *Soft Matter* **2013**, *9*, 2047–2055. [CrossRef]

31. You, J.; Xiang, M.; Hu, H.; Cai, J.; Zhou, J.; Zhang, Y. Aqueous synthesis of silver nanoparticles stabilized by cationic cellulose and their catalytic and antibacterial activities. *RSC Advances* **2013**, *3*, 19319–19329. [CrossRef]

32. Li, G.B.; Fu, Y.J.; Shao, Z.Y.; Zhang, F.S.; Qin, M.H. Preparing cationic cellulose derivative in naoh/urea aqueous solution and its performance as filler modifier. *Bioresources* **2015**, *10*, 7782–7794. [CrossRef]

33. Song, Y.; Zhang, L.; Gan, W.; Zhou, J.; Zhang, L. Self-assembled micelles based on hydrophobically modified quaternized cellulose for drug delivery. *Colloids Surf. B* **2011**, *83*, 313–320. [CrossRef] [PubMed]

34. Kalashnikova, I.; Bizot, H.; Cathala, B.; Capron, I. New pickering emulsions stabilized by bacterial cellulose nanocrystals. *Langmuir* **2011**, *27*, 7471–7479. [CrossRef] [PubMed]

35. Andresen, M.; Stenius, P. Water-in-oil emulsions stabilized by hydrophobized microfibrillated cellulose. *J. Disper. Sci. Technol.* **2007**, *28*, 837–844. [CrossRef]

36. Okushita, K.; Komatsu, T.; Chikayama, E.; Kikuchi, J. Statistical approach for solid-state nmr spectra of cellulose derived from a series of variable parameters. *Polym. J.* **2012**, *44*, 895–900. [CrossRef]

37. Chaker, A.; Boufi, S. Cationic nanofibrillar cellulose with high antibacterial properties. *Carbohyd. Polym.* **2015**, *131*, 224–232. [CrossRef] [PubMed]

38. Song, Y.; Sun, Y.; Zhang, X.; Zhou, J.; Zhang, L. Homogeneous quaternization of cellulose in NaOH/urea aqueous solutions as gene carriers. *Biomacromolecules* **2008**, *9*, 2259–2264. [CrossRef] [PubMed]

39. Yan, L.; Tao, H.; Bangal, P.R. Synthesis and flocculation behavior of cationic cellulose prepared in a naoh/urea aqueous solution. *CLEAN Soil Air Water* **2009**, *37*, 39–44. [CrossRef]

40. Stenstad, P.; Andresen, M.; Tanem, B.S.; Stenius, P. Chemical surface modifications of microfibrillated cellulose. *Cellulose* **2008**, *15*, 35–45. [CrossRef]

41. Dzidic, I.; Somerville, A.C.; Raia, J.C.; Hart, H.V. Determination of naphthenic acids in california crudes and refinery wastewaters by fluoride ion chemical ionization mass spectrometry. *Anal. Chem.* **1988**, *60*, 1318–1323. [CrossRef]

42. Fan, T.P. Characterization of naphthenic acids in petroleum by fast atom bombardment mass spectrometry. *Energy Fuels* **1991**, *5*, 371–375. [CrossRef]

43. Wong, D.C.L.; van Compernolle, R.; Nowlin, J.G.; O'Neal, D.L.; Johnson, G.M. Use of supercritical fluid extraction and fast ion bombardment mass spectrometry to identify toxic chemicals from a refinery effluent adsorbed onto granular activated carbon. *Chemosphere* **1996**, *32*, 1669–1679. [CrossRef]

44. St. John, W.P.; Rughani, J.; Green, S.A.; McGinnis, G.D. Analysis and characterization of naphthenic acids by gas chromatography-electron impact mass spectrometry of tert.-butyldimethylsilyl derivatives. *J. Chromatogr. A* **1998**, *807*, 241–251. [CrossRef]

45. Hsu, C.S.; Dechert, G.J.; Robbins, W.K.; Fukuda, E.K. Naphthenic acids in crude oils characterized by mass spectrometry. *Energy Fuels* **1999**, *14*, 217–223. [CrossRef]

46. Azad, F.S.; Abedi, J.; Iranmanesh, S. Removal of naphthenic acids using adsorption process and the effect of the addition of salt. *J. Environ. Sci. Health Part A* **2013**, *48*, 1649–1654. [CrossRef] [PubMed]

47. Iranmanesh, S.; Harding, T.; Abedi, J.; Seyedeyn-Azad, F.; Layzell, D.B. Adsorption of naphthenic acids on high surface area activated carbons. *J. Environ. Sci. Heal. A* **2014**, *49*, 913–922. [CrossRef] [PubMed]

48. Kushwaha, A.K.; Gupta, N.; Chattopadhyaya, M.C. Removal of cationic methylene blue and malachite green dyes from aqueous solution by waste materials of daucus carota. *J. Saudi Chem. Soc.* **2014**, *18*, 200–207. [CrossRef]

49. Marshall, A.G.; Rodgers, R.P. Petroleomics: Chemistry of the underworld. *Proc. Natl. Acad. Sci. USA* **2008**, *105*, 18090–18095. [CrossRef] [PubMed]

50. Mohamed, M.H.; Wilson, L.D.; Shah, J.R.; Bailey, J.; Peru, K.M.; Headley, J.V. A novel solid-state fractionation of naphthenic acid fraction components from oil sands process-affected water. *Chemosphere* **2015**, *136*, 252–258. [CrossRef] [PubMed]

51. Mufazzal Saeed, M.; Ahmed, M. Effect of temperature on kinetics and adsorption profile of endothermic chemisorption process: –tm(iii)–pan loaded puf system. *Separ. Sci. Technol.* **2006**, *41*, 705–722. [CrossRef]

52. Saha, P.; Chowdhury, S. *Insight into Adsorption Thermodynamics, Thermodynamics*; Tadashi, M., Ed.; InTech: Rijeka, Croatia, 2011; pp. 349–365.

53. Inglezakis, V.J.; Zorpas, A.A. Heat of adsorption, adsorption energy and activation energy in adsorption and ion exchange systems. *Desalin. Water Treat.* **2012**, *39*, 149–157. [CrossRef]

54. Langmuir, I. The adsorption of gases on plane surfaces of glass, mica and platinum. *J. Am. Chem. Soc.* **1918**, *40*, 1361–1403. [CrossRef]

55. Freundlich, H.M.F. Over the adsorption in solution. *J. Phys. Chem.* **1906**, *57A*, 385–470.

56. Sips, R. Structure of a catalyst surface. *J. Chem. Phys.* **1948**, *16*, 490–495. [CrossRef]

57. Mohamed, M.; Wilson, L. Kinetic uptake studies of powdered materials in solution. *Nanomaterials* **2015**, *5*, 969–980. [CrossRef]

Article

Pickering Emulsion-Based Marbles for Cellular Capsules

Guangzhao Zhang and Chaoyang Wang *

Research Institute of Materials Science, South China University of Technology, Guangzhou 510640, China; msgzhzhang@mail.scut.edu.cn
* Correspondence: zhywang@scut.edu.cn; Tel.: +86-20-2223-6269

Academic Editor: To Ngai
Received: 27 June 2016; Accepted: 7 July 2016; Published: 14 July 2016

Abstract: The biodegradable cellular capsule, being prepared from simple vaporization of liquid marbles, is an ideal vehicle for the potential application of drug encapsulation and release. This paper reports the fabrication of cellular capsules via facile vaporization of Pickering emulsion marbles in an ambient atmosphere. Stable Pickering emulsion (water in oil) was prepared while utilizing dichloromethane (containing poly(L-lactic acid)) and partially hydrophobic silica particles as oil phase and stabilizing agents respectively. Then, the Pickering emulsion marbles were formed by dropping emulsion into a petri dish containing silica particles with a syringe followed by rolling. The cellular capsules were finally obtained after the complete vaporization of both oil and water phases. The technique of scanning electron microscope (SEM) was employed to research the microstructure and surface morphology of the prepared capsules and the results showed the cellular structure as expected. An in vitro drug release test was implemented which showed a sustained release property of the prepared cellular capsules. In addition, the use of biodegradable poly(L-lactic acid) and the biocompatible silica particles also made the fabricated cellular capsules of great potential in the application of sustained drug release.

Keywords: Pickering emulsion; liquid marbles; cellular capsule; drug release; biodegradable

1. Introduction

It is well known that the term 'liquid marble' is defined to describe aqueous droplets which are enwrapped by self-organized hydrophobic powders [1,2]. Since Aussillous transported liquid marble via rolling water droplets across hydrophobic grains [3], a large number of researchers have focused on the investigation of liquid marbles [4–6]. In fact, the liquid marbles had already existed for a long time, if one considers rainwater dropping into dry dirt or the water droplets falling into large quantity of wheat flour, although few people had previously paid attention to it. Due to their distinctive properties—including the ability to be divided or fused together with self-recovery enwrapped layers [7], low frictional resistance derived from small a contact area with the subsurface [8] as well as relatively facile manipulation—extensive research on liquid marbles has been carried out recently. Moreover, these unique properties make investigation of liquid marbles not limited only to theoretical research, but also extended to some practical applications, such as miniature or micro-chemical reactors [9], sensors [10], pollution detection [11], vehicles for transporting microfluidics [12], and oil spill treatment [13]. For instance, Arbatan and coworkers employed the hydrophobic powder precipitated calcium carbonate (PCC) to prepare the 'blood liquid marbles' as micro-reactors for biological reactions and diagnostic experiments [14]. Interestingly, the blood marbles they reported could identify the blood type rapidly while showing some advantages—including low cost, which free medical facilities rely on, and disposability—when compared with conventional blood typing techniques. Another interesting example worthy of mention is Fujii et al.'s work [15]: They reported a smart liquid marble which could move on the surface of water under the control of near-infrared laser or sunlight. Moreover, the marbles they prepared could also release the encapsulated material at a specific time and place by controllable external stimuli, making it an ideal candidate for target controlled release. In addition, other explorations of liquid marbles for chemical synthesis, mass transport and self-assembly have been reported recently [16], suggesting promising potential applications of liquid marbles.

Similar to, but different from liquid marbles, Pickering emulsion also makes use of solid particles to absorb to a liquid–liquid interface, which leads to the formation of emulsion and sequentially keeps it stable. The biggest difference lies in the medium surrounding the emulsion droplets being a liquid phase (water or oil), while the encapsulating medium is replaced by air for liquid marbles [17]. The amphiphilic solid particles used for the stabilization of Pickering emulsion exhibit low toxicity and favorable biocompatibility when compared with traditional emulsion stabilized with harmful surfactant, which makes the Pickering emulsion an environmentally friendly candidate for applications in the fields of food [18], cosmetics [19], and green catalysis [20]. In view of the flexibility in the selection of solid particles and oil phase, there is nothing surprising about the fact that Pickering emulsion has drawn much attention during the last decades. Additionally, the unique property of Pickering emulsion templates was also explored fully by researchers to prepare porous material [21] or cellular structure [22], further expanding its practical applications.

Inspired by the pioneer's work, we combined the merits of liquid marbles and Pickering emulsion in this report, and prepared the Pickering emulsion-based marbles to fabricate cellular capsules via a facile rolling method. To date, although a considerable amount of research on liquid marbles has been undertaken, examples of exploiting Pickering emulsion as an encapsulated liquid to prepare emulsion marbles are rare. The design of our work is presented in detail in Scheme 1: The water in oil Pickering emulsion was prepared first using dichloromethane (containing poly(L-lactic acid)) as the oil phase; then the fabricated emulsion was transferred drop by drop into a petri dish containing sufficient solid particles, using a syringe equipped with a 26 gauge needle; followed by immediate rolling to form stable emulsion-based marbles. The fabricated emulsion marbles were transported to a glass substrate and then dried in air at room temperature. The cellular capsules were obtained after the complete evaporation of both dichloromethane and water. Influences on the formation of emulsion marbles, including the concentration of polymer, stabilizer particle content and internal phase volume fraction were studied amply, and the technique of scanning electron microscopy (SEM) was also utilized to characterize the microstructure and morphology of the prepared capsules. Moreover, an in vitro drug release test was also carried out and the results suggest that the fabricated capsule is a candidate for sustained drug release.

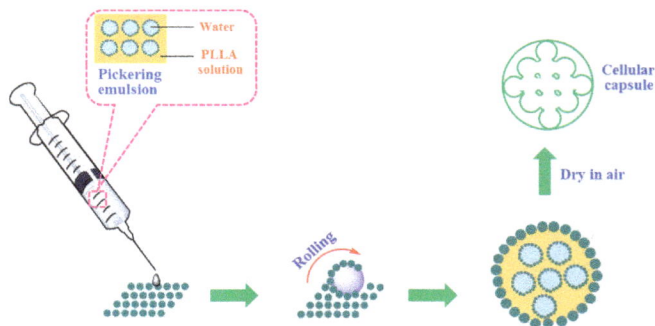

Scheme 1. The schematic illustration of the preparation of Pickering emulsion-based marbles and cellular capsules.

2. Materials and Methods

2.1. Materials

Poly(L-lactic acid) (PLLA, molecular weight M_w = 200,000 g/mol) was purchased from Shandong Medical Instrument Research Institute (Jinan, China). Partially hydrophobic silica (H$_{30}$) (with an average diameter of 20 nm and 50% of surface hydroxyl content) was obtained from Wacker Chemie Company (Munich, Germany). Dichloromethane (CH$_2$Cl$_2$) was purchased from Guangzhou Chemical Factory (Guangzhou, China). Enrofloxacin (98%) was bought from J&K Scientific Co., Ltd. (Shanghai, China). Water used in all experiments was purified with a Millipore purification apparatus (Boston, MA, USA) with a resistance higher than 18.0 M$\Omega \cdot$ cm.

2.2. Preparation of Pickering Emulsion

The water in oil Pickering emulsion was prepared according to the recipes listed in Table 1. Typically, a certain amount of PLLA was dissolved in dichloromethane (CH$_2$Cl$_2$) to form a homogeneous PLLA solution. Then, the partially hydrophobic silica particles (H$_{30}$) were uniformly dispersed into PLLA solution with the assistance of sonication. After that, ultrapure water was added to the dispersion followed by emulsifying with a vortex mixer (Fisher Scientific, Haimen, China) at 3000 rpm for 3 min. It was noted that the sample P$_{0-0.1}$ was just a comparison which was prepared by dissolving 0.10 g of PLLA into 2 mL of CH$_2$Cl$_2$.

Table 1. Recipes of the different Pickering emulsion.

Serial	H$_{30}$ (w/v %) [a]	PLLA (w/v %) [a]	CH$_2$Cl$_2$ (mL)	Water (mL)
P$_{0-0.1}$	0	5	2	0
P$_{0.01-0.1-2}$	0.5	5	2	2
P$_{0.02-0.1-2}$	1.0	5	2	2
P$_{0.04-0.1-2}$	2.0	5	2	2
P$_{0.04-0.12-2}$	2.0	6	2	2
P$_{0.04-0.1-3}$	2.0	5	2	3

[a] With respect to oil phase.

2.3. Preparation of Pickering Emulsion-Based Marbles and Cellular Capsules

The general preparation method of Pickering emulsion-based marbles is described as follows. An adequate amount of as-prepared Pickering emulsion was loaded to a 1 mL syringe equipped with a 26-gauge needle at first. Then, the loaded emulsion was injected dropwise into a petri dish containing H$_{30}$ silica powder followed by vigorously shaking the dish in a horizontal direction, keeping the emulsion droplet rolling until its surface was totally covered by silica particles. The Pickering emulsion marbles were obtained after removing residual silica powder. For the fabrication of cellular capsules, the only procedure which needed to be implemented was to place the prepared Pickering emulsion marbles in air at an ambient temperature for 48 h for the complete vaporization of both oil (CH$_2$Cl$_2$) and water phases.

2.4. In Vitro Drug Release

To fabricate the drug-loaded cellular structure, 1 mg of enrofloxacin was dissolved into 2 mL of CH$_2$Cl$_2$ solution alongside the PLLA prior to adding H$_{30}$ particles, and the following procedure was identical to the description in Section 2.3 above. The in vitro drug release investigation was implemented by measuring the absorbance of released enrofloxacin at 271 nm with an Ultraviolet–visible (UV–vis) spectrophotometer (U-3010, Hitachi, Tokyo, Japan). For this experiment, typically 20 mg (containing 200 μg of enrofloxacin) of cellular capsule was placed into a glass bottle containing 20 mL of ultrapure water. Then, the bottle was incubated in a shaking incubator at 37 °C for release. 2 mL of release medium was withdrawn and equal fresh ultrapure water was added as a substitute at predetermined time intervals. The collected aqueous medium was analyzed by a UV–vis technique according to the calibration curves of enrofloxacin established from standard enrofloxacin solution. The assay was performed in triplicate.

2.5. Characterization

The Pickering emulsion droplets were observed with an optical microscope (Carl Zeiss, Oberkochen, Germany) equipped with a digital camera. The morphology of cellular capsule was characterized with a Zeiss EVO 18 scanning electron microscope (SEM) (Oberkochen, Germany). Generally, capsule samples were stuck onto conductive tape prior to gold coating with a sputter coater, followed by observation under a 10 kV acceleration voltage.

2.6. Estimation of H_{30} Particles Content Using for Stabilizing Emulsion Marbles

In order to estimate the content of H_{30} particles which serve to stabilize emulsion marbles, we set the volume of each Pickering emulsion droplet as 20 µL and measured the mass of ten dried capsules which was 11.6 mg. Then, 200 µL of Pickering emulsion was placed in a glass bottle in an ambient atmosphere for the complete vaporization of CH_2Cl_2 and water. After that, the mass of the residual scaffold was also measured and the result was 10.8 mg. Therefore, the average content (W_a) of H_{30} particles used for stabilizing emulsion marbles was:

$$W_a = (11.6 - 10.8)/10 = 0.08 \, \text{mg}$$

3. Result and Discussion

3.1. Preparation of Pickering Emulsion

The Pickering emulsion was prepared via simple emulsifying of the mixture of PLLA solution and water with the help of H_{30} stabilizer. It was observed that the concentration of H_{30} silica particles have a significant influence on the stability of prepared Pickering emulsion. We set the volume of oil and water phase at 2 mL respectively and varied the amount of H_{30} silica particles for detailed research in our experiments (Table 1). It was found that when the concentration of H_{30} particles was 0.5 w/v % (sample $P_{0.01-0.1-2}$, with respect of oil phase, and similarly from here), stratification appeared in the formed Pickering emulsion after it was placed in an ambient atmosphere for 48 h, which suggested that 0.5 w/v % of H_{30} silica particles was not enough for the formation of a stable Pickering emulsion. In fact, it was too difficult to observe the morphology of these Pickering emulsion droplets with optical microscope in conjunction with the volatility of CH_2Cl_2. However, when the concentration of H_{30} particles were increased to 1 w/v % and 2 w/v % ($P_{0.02-0.1-2}$ and $P_{0.04-0.1-2}$), the formed emulsion remained stable for 2 days. An optical microscope was employed for the observation of formed Pickering emulsion droplets, and the result is shown in Figure 1. As an intuitive demonstration, the droplets' size distribution was also analyzed and the results are exhibited in Figure S1. It was clear that the size of the emulsion droplets decreased dramatically with the increase of H_{30} stabilizer, which accorded with the normal emulsion stabilization mechanism. The concentration of PLLA in CH_2Cl_2 did not show significant influence on the stability of emulsion in our experiments, but did increase the viscosity of formed emulsion. When raising the concentration of PLLA from 5 w/v % to 6 w/v % ($P_{0.04-0.5-2}$), the viscosity of emulsion increased apparently and became difficult to inject dropwise by syringe (Figure S2A). It should be noted that the increase of internal volume fraction ($P_{0.04-0.5-3}$) also made the formed emulsion more viscous, which resulted in the same phenomenon as the increment of PLLA. To determine the type of Pickering emulsion, we carried out a simple test by dropping the formed emulsion into a CH_2Cl_2 solution and water respectively. As shown in Figure S3, the emulsion droplets were destroyed and spread out in CH_2Cl_2, while remaining stable in water, indicating that the formed emulsion was water in oil type (W/O).

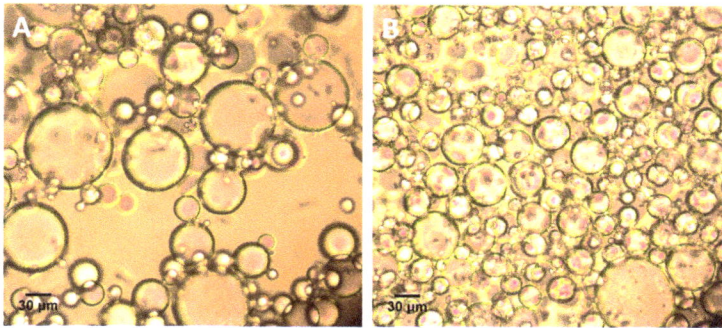

Figure 1. Optical microscope image of prepared Pickering emulsion with different H_{30} concentrations: (**A**) 1 w/v % of H_{30} silica particles, sample $P_{0.02-0.1-2}$; (**B**) 2 w/v % of H_{30} silica particles, sample $P_{0.04-0.1-2}$.

3.2. Preparation of Pickering Emulsion Marbles and Cellular Capsule

It is well known that the adhesivity of powder to a specific liquid is vital for the formation of stable liquid marbles. Considering the fact that the Pickering emulsion was water in oil type, partially hydrophobic silica powder (H_{30}) was chosen to cover the surface of Pickering emulsion droplets for the preparation of emulsion marbles. It was found that the stability of Pickering emulsion was of great significance to the formation of emulsion marbles. When the concentration of stabilizer particle (H_{30}) was 0.5 w/v %, the formed emulsion marbles remained intact in a petri dish containing a large amount of silica powder, but broke up immediately after transfer to a glass base (Figure S2C). This phenomenon suggested that the Pickering emulsion in an unstable state was difficult to cover with powder for forming stable emulsion marbles. In fact, the unstable Pickering emulsion should separate to form the initial oil and water phase from a uniform state according to the Ostwald Ripening theory, resulting in direct contact of external H_{30} powder with split water. Due to the hydrophobic property of H_{30} silica powder, it was difficult to adhere it to the surface of a water droplet for the formation of stable liquid marbles. In order to confirm our speculation, an experiment using pure water to prepare liquid marbles with H_{30} was carried out and the results are shown in Figure S4. Obviously, although the marble remained stable in large qualities of H_{30} powders, it collapsed immediately after transfer onto a glass subtrate. When the concentration of H_{30} particles was increased to 1 w/v %, the formed emulsion marbles became more stable than at 0.5 w/v %. However, an unexpected phenomenon appeared—the water droplet was squeezed out from one side of the marble (Figure S2D,E) during the drying process. Considering the fact that the evaporation rate of CH_2Cl_2 was much faster than that of water at room temperature, partial demulsification appeared in the formed marble during the vaporization of oil phase, resulting in the extrusion of the split water droplet from one side of the marble. When increasing the concentration of particles to 2 w/v %, the Pickering emulsion formed spherical and stable marbles (Figure 2B), and the formed emulsion marbles did not exhibit obvious shrinkage after the complete evaporation of both oil and water phases. Three reasons were proposed to explain the successful fabrication of Pickering emulsion marbles: (1) The increment of stabilizer particles made the emulsion much more stable, which inhibited the aggregation of water droplets during the vaporization of CH_2Cl_2; (2) The increment of stabilizer decreased the size of emulsion droplets, slowing down the evaporation of CH_2Cl_2; (3) The dissolved PLLA would separate out and sequentially adhere to the surface of emulsion droplets, impeding the accumulation of a disperse phase. It must be pointed out that excessive PLLA in CH_2Cl_2 solution ($P_{0.04-0.12-2}$) or exorbitant internal phase volume fraction ($P_{0.04-0.1-3}$) would make the formed Pickering emulsion too viscous to be injected drop by drop from the needle (Figure S2A), which is adverse for the fabrication of spherical emulsion marbles.

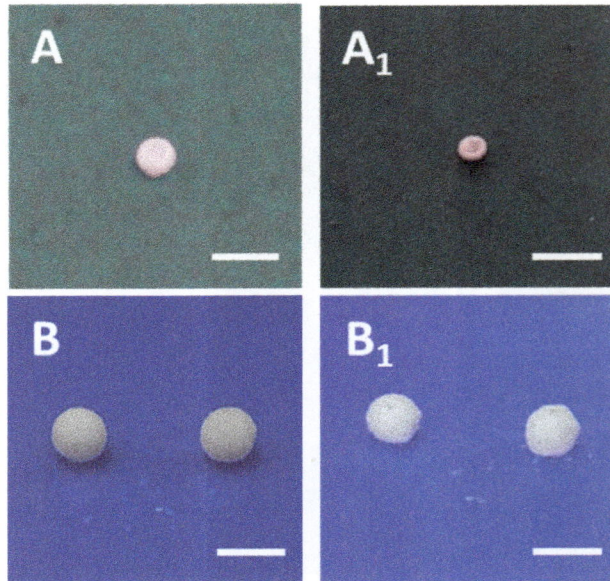

Figure 2. Digital photographs of prepared liquid marbles: (**A**) liquid marble directly deriving from PLLA solution; (**A$_1$**) the formed capsule after the vaporization of dichloromethane; (**B**) Pickering emulsion marbles; (**B$_1$**) the formed cellular capsule after the vaporization of both oil and water phase. All scale bars show 3 mm.

Pure CH_2Cl_2 liquid marbles containing PLLA were also prepared, for comparison (Figure 2A,A$_1$). It was clear that the CH_2Cl_2 liquid marble underwent severe shrinkage after drying while Pickering emulsion marbles kept their initial shape (Figure 2B,B$_1$). This phenomenon demonstrated that the Pickering emulsion marbles successfully inhibited the shrinkage of PLLA film during the solution vaporization. In addition, the size of Pickering emulsion marbles was much larger than that of CH_2Cl_2 liquid marbles, which was attributed to the increasing viscosity of emulsion compared with CH_2Cl_2 solution. It should be noted that the Pickering emulsion droplet was rolled in a petri dish containing H$_{30}$ silica particles until its surface was completely covered with powder (Figure S2B). The average amount of silica particles in one capsule was estimated to be 0.08 mg.

The technique of SEM was employed for interrogating the microstructure and morphology of the prepared cellular capsules deriving from Pickering emulsion marbles. Figure 3A shows the overall morphology of capsules. Holes with a diameter of approximately 80–100 µm were distributed on the capsule surface. Figure S5 helps to explain this phenomenon: During the drying process, the CH_2Cl_2 liquid layer between Pickering emulsion droplets (especially for large droplets near the marble surface) and marble surface was not thick enough, and the separated PLLA film after the complete vaporization of CH_2Cl_2 was too thin to cover the whole surface of the droplets, resulting in the generation of holes on the surface of the capsule. In addition, considering that the evaporation rate of CH_2Cl_2 was much faster than that of water, the water droplets still existed after the complete vaporization of CH_2Cl_2, and the surface tension from those droplets facilitated the formation of holes. Moreover, the diameter of the hole was about 50 µm, and this value was identical to the size of some Pickering emulsion droplets in Figure 1B, which supported the proposed explanation. We found a cellular structure inside of the capsule (Figure 3B), which was identical to our expectation. To further confirm the internal structure, we cut the capsule into slices and characterized them with SEM (Figure 3D). It was obvious that the inner morphology of the capsule was a cellular structure, which was consistent with the conclusion coming from Figure 3B. Figure 3C is the local enlarged image of the capsule surface. It was clear that the capsule surface was a porous morphology at high magnification, though they were relatively flat at low magnification. This phenomenon could also be attributed to Pickering emulsion templates.

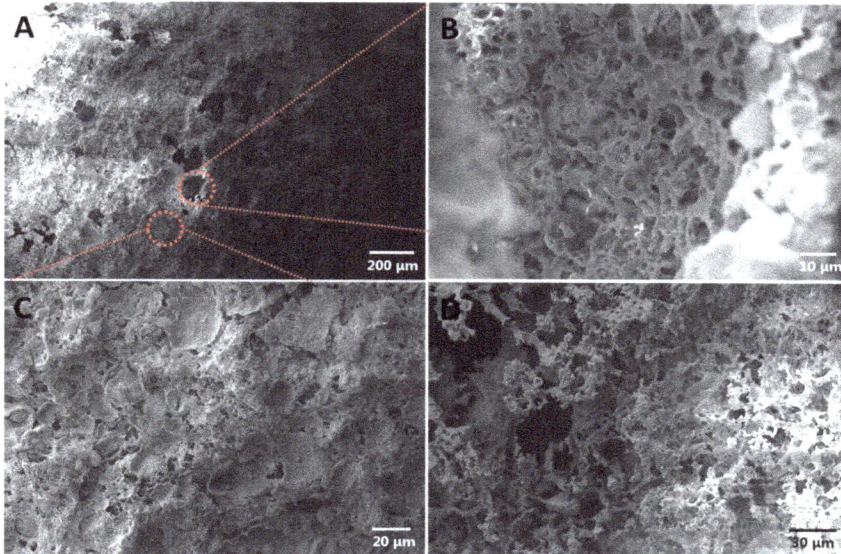

Figure 3. SEM images of prepared cellular capsule derived from Pickering emulsion-based marbles ($P_{0.02\text{-}0.1\text{-}2}$): (**A**) Overall morphology of prepared cellular capsules; (**B**) Internal cellular structure of the capsule from the perspective of large hole distributing on the surface; (**C**) Surface morphology of the capsule; (**D**) Inner structure of the capsule after incision.

3.3. In Vitro Drug Release

Many substances, such as hydrogels [23] and particles [24], have been employed as vehicles for the encapsulation and sustained release of the drug in biological territory. The porous PLLA scaffold [25], which has proved to be a preferred carrier for sustained drug release inspired us to attempt a similar application of the prepared cellular capsules. The enrofloxacin, a common bactericide for animals, was exploited for exploration of the sustained drug release of prepared cellular capsules. Figure 4 shows the in vitro release curve of enrofloxacin encapsulated in capsule scaffold. It can be seen that the enrofloxacin was released relatively quickly during the initial period of 50 h, and the release percentage was up to 36.0%. After this period, the release rate of enrofloxacin from capsules became slow until the release percent reached a stable value (about 42%). After 100 h, the release curve did not show any more apparent fluctuation, suggesting the drug release became steady with a final value of 42%. In conjunction with the biocompatibility of PLLA and H_{30} silica particles, the prepared cellular capsules could be a promising candidate for the application of sustained drug release.

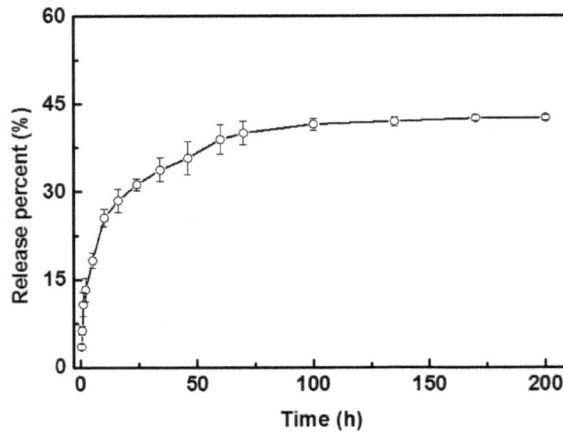

Figure 4. In vitro release curve of cellular capsule deriving Pickering emulsion marbles.

4. Conclusions

In this report, we prepared Pickering emulsion-based marbles and fabricated the cellular capsules from the prepared emulsion marbles via facile vaporization of both oil and water phases in ambient atmosphere. The solution of PLLA in CH_2Cl_2 was utilized as the oil phase of Pickering emulsion, and H_{30} silica particles served as both stabilizer for Pickering emulsion and enwrapping powder for emulsion marbles. Influences on the formation of emulsion marbles were explored in detail and it was concluded that favorable stability and appropriate viscosity of the emulsion would facilitate the fabrication of emulsion marbles. The morphology of the obtained cellular capsules was characterized by SEM and the result was identical to our expectation. In addition, the test of in vitro drug release using enrofloxacin as a model showed that the fabricated capsules impart favorable property of sustained drug release, which make it a potential candidate in the application of biomedical territory combining the excellent biocompatibility and biodegradability of PLLA.

Supplementary Materials: The supplementary materials are available online at www.mdpi.com/1996-1944/9/7/572/s1.

Acknowledgments: The authors acknowledge financial support from the National Natural Science Foundation of China (21474032).

Author Contributions: Chaoyang Wang designed and proposed the exploration scheme, and Guangzhao Zhang performed the experiments and wrote the paper.

Conflicts of Interest: The authors declare no conflict of interest.

References

1. Shuttle, D.; Schmitz, F.T.; Brunner, H. Predominantly Aqueous Compositions in a Fluffy Powdery form Approximating Powdered Solids Behavior and Process for Forming Same. U.S. Patent 3393155A, 16 July 1968.
2. Aussillous, P.; Quéré, D. Properties of liquid marbles. *Proc. R. Soc. A* **2006**, *462*, 973–999. [CrossRef]
3. Aussillous, P.; Quéré, D. Liquid marbles. *Nature* **2001**, *411*, 924–927. [CrossRef] [PubMed]
4. Mchale, G.; Newton, M.I. Liquid marbles: Topical context within soft matter and recent progress. *Soft Matter* **2015**, *11*, 2530–2546. [CrossRef] [PubMed]
5. Sheng, Y.F.; Sun, G.Q.; Wu, J.; Ma, G.H.; Ngai, T. Silica-Based Liquid Marbles as Microreactors for the Silver Mirror Reaction. *Angew. Chem. Int. Ed.* **2015**, *54*, 7012–7017. [CrossRef] [PubMed]
6. Yildirim, A.; Budunoglu, H.; Daglar, B.; Deniz, H.; Bayindir, M. One-Pot Preparation of Fluorinated Mesoporous Silica Nanoparticles for Liquid Marble Formation and Superhydrophobic Surfaces. *ACS Appl. Mater. Interfaces* **2011**, *3*, 1804–1808. [CrossRef] [PubMed]

7. Mchale, G.; Newton, M.I. Liquid marbles: Principles and applications. *Soft Matter* **2011**, *7*, 5473–5481. [CrossRef]

8. Sivan, V.; Tang, S.Y.; O'Mullane, A.P.; Petersen, P.; Eshtiaghi, N.; Kalantar-zadeh, K.; Mitchell, A. Liquid Metal Marbles. *Adv. Funct. Mater.* **2013**, *23*, 144–152. [CrossRef]

9. Xue, Y.H.; Wang, H.X.; Zhao, Y.; Dai, L.M.; Feng, L.F.; Wang, X.G.; Lin, T. Magnetic Liquid Marbles: A "Precise" Miniature Reactor. *Adv. Mater.* **2010**, *22*, 4814–4818. [CrossRef] [PubMed]

10. Fujii, S.; Kameyama, S.; Armes, S.P.; Dupin, D.; Suzaki, M.; Nakamura, Y. pH-responsive liquid marbles stabilized with poly(2-vinylpyridine) particles. *Soft Matter* **2010**, *6*, 635–640. [CrossRef]

11. Han, X.M.; Lee, H.K.; Lee, Y.H.; Hao, W.; Liu, Y.J.; Phang, I.Y.; Li, S.Z.; Ling, X.Y. Identifying Enclosed Chemical Reaction and Dynamics at the Molecular Level Using Shell-Isolated Miniaturized Plasmonic Liquid Marble. *J. Phys. Chem. Lett.* **2016**, *7*, 1501–1506. [CrossRef] [PubMed]

12. Samiei, E.; Tabrizian, M.; Hoorfar, M. A review of digital microfluidics as portable platforms for lab-on a-chip applications. *Lab Chip* **2016**, *16*, 2376–2396. [CrossRef] [PubMed]

13. Huang, S.Y.; Zhang, Y.; Shi, J.F.; Huang, W.P. Superhydrophobic particles derived from nature-inspired polyphenol chemistry for liquid marble formation and oil spills treatment. *ACS Sustain. Chem. Eng.* **2016**, *4*, 676–681. [CrossRef]

14. Arbatan, T.; Li, L.Z.; Tian, J.F.; Shen, W. Liquid Marbles as Micro-bioreactors for Rapid Blood Typing. *Adv. Healthc. Mater.* **2012**, *1*, 80–83. [CrossRef] [PubMed]

15. Paven, M.; Mayama, H.; Sekido, T.; Butt, H.J.; Nakamura, Y.; Fujii, S. Light-Driven Delivery and Release of Materials Using Liquid Marbles. *Adv. Funct. Mater.* **2016**, *26*, 3199–3206. [CrossRef]

16. Chu, Y.; Wang, Z.K.; Pan, Q.M. Constructing Robust Liquid Marbles for Miniaturized Synthesis of Graphene/Ag Nanocomposite. *ACS Appl. Mater. Interfaces* **2014**, *6*, 8378–8386. [CrossRef] [PubMed]

17. Destribats, M.; Rouvet, M.; Gehin-Delval, C.; Schmitt, C.; Binks, B.P. Emulsions stabilised by whey protein microgel particles: Towards food-grade Pickering emulsions. *Soft Matter* **2014**, *10*, 6941–6954. [CrossRef] [PubMed]

18. Tan, H.; Sun, G.Q.; Lin, W.; Mu, C.D.; Ngai, T. Gelatin Particle-Stabilized High Internal Phase Emulsions as Nutraceutical Containers. *ACS Appl. Mater. Interfaces* **2014**, *6*, 13977–13984. [CrossRef] [PubMed]

19. Sarkar, A.; Murray, B.; Holmes, M.; Ettelaie, R.; Abdalla, A.; Yang, X.Y. In vitro digestion of Pickering emulsions stabilized by soft whey protein microgel particles: Influence of thermal treatment. *Soft Matter* **2016**, *12*, 3558–3569. [CrossRef] [PubMed]

20. Yu, Y.H.; Fu, L.M.; Zhang, F.W.; Zhou, T.; Yang, H.Q. Pickering-Emulsion Inversion Strategy for Separating and Recycling Nanoparticle Catalysts. *Chem. Phys. Chem.* **2014**, *15*, 841–848. [CrossRef] [PubMed]

21. Zou, S.W.; Hu, Y.; Wang, C.Y. One-Pot Fabrication of Rattle-Like Capsules with Multicores by Pickering Based Polymerization with Nanoparticle Nucleation. *Macromol. Rapid Commun.* **2014**, *35*, 1414–1418. [CrossRef] [PubMed]

22. Hu, Y.; Ma, S.S.; Yang, Z.H.; Zhou, W.Y.; Du, Z.S.; Huang, J.; Yi, H.; Wang, C.Y. Facile fabrication of poly(L-lactic acid) microsphere-incorporated calcium alginate/hydroxyapatite porous scaffolds based on Pickering emulsion templates. *Coll. Surf. B Biointerfaces* **2016**, *140*, 382–391. [CrossRef] [PubMed]

23. Appel, A.E.; Tibbitt, M.W.; Webber, M.J.; Mattix, B.A.; Veiseh, O.; Langer, R. Self-assembled hydrogels utilizing polymer–nanoparticle interactions. *Nat. Commun.* **2015**, *6*, 6295–6313. [CrossRef] [PubMed]

24. Dou, Y.; Guo, J.W.; Chen, X.; Han, S.L.; Xu, X.Q.; Shi, Q.; Jia, Y.; Liu, Y.; Deng, Y.C.; Zhang, J.X.; et al. Sustained delivery by a cyclodextrin material-based nanocarrier potentiates antiatherosclerotic activity of rapamycin via selectively inhibiting mTORC1 in mice. *J. Control. Release* **2016**, *235*, 48–62. [CrossRef] [PubMed]

25. Hu, Y.; Gu, X.Y.; Yang, Y.; Huang, J.; Hu, M.; Chen, W.K.; Tong, Z.; Wang, C.Y. Facile Fabrication of Poly(L-lactic Acid)-Grafted Hydroxyapatite/Poly(lactic-co-glycolic Acid) Scaffolds by Pickering High Internal. *ACS Appl. Mater. Interfaces* **2014**, *6*, 17166–17175. [CrossRef] [PubMed]

MDPI AG

St. Alban-Anlage 66

4052 Basel, Switzerland

Tel. +41 61 683 77 34

Fax +41 61 302 89 18

http://www.mdpi.com

Materials Editorial Office

E-mail: materials@mdpi.com

http://www.mdpi.com/journal/materials